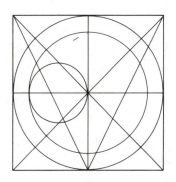

OSBORNE/McGraw-Hill

SCIENCE AND ENGINEERING PROGRAMS   APPLE II®Edition

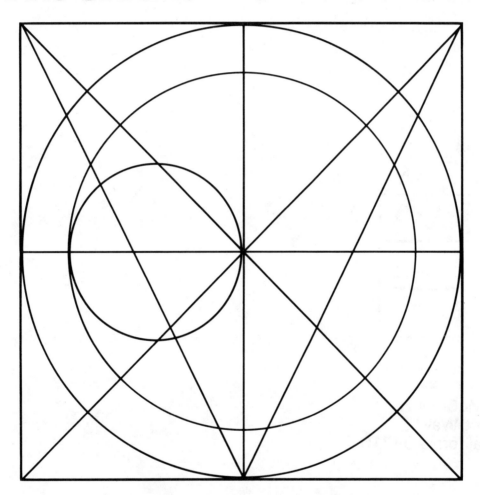

Edited by John Heilborn

Contributors:

Richard Bennington	William Harlow
Marcial Blondet	Shafi Motiwalla
Kenneth Douglass	Juan Carlos Simo

DISCLAIMER OF WARRANTIES
AND LIMITATION OF LIABILITIES

The authors have taken due care in preparing this book and the programs in it, including research, development, and testing to ascertain their effectiveness. The authors and publishers make no expressed or implied warranty of any kind with regard to these programs or the supplementary documentation in this book. In no event shall the authors or publishers be liable for incidental or consequential damages in connection with or arising from the furnishing, performance, or use of any of these programs.

Published by
OSBORNE/McGraw-Hill
630 Bancroft Way
Berkeley, California 94710
U.S.A.

For information on translations and book distributors outside of the U.S.A., please write OSBORNE/McGraw-Hill.

SCIENCE AND ENGINEERING PROGRAMS, APPLE II® EDITION

123456789 87654321

ISBN 0-931900-63-2

Acknowledgments

No book of this sort can be the work of only one person. The technical advisor who helped evaluate and select most of the programs was Shafi Motiwalla. Martin McNiff provided the technical support needed in the initial phases, including assembling the original program titles for publication.

The authors who wrote the programs are: Richard Bennington, Marcial Blondet, Kenneth Douglass, Professor William Harlow, and Juan Carlos Simo. Their work speaks for itself.

Finally, we wish to thank Cynthia Greever for her work in running and in many cases rerunning the programs to ensure they got to the book in good order.

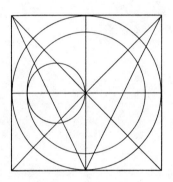

Introduction

The programs in this book were tested and run on an Apple II[1] microcomputer. None requires more than 6K of RAM memory space to enter, although it may be necessary to have more RAM for data and/or arrays or matrices that are generated by the programs themselves.

Each program includes a brief discussion of its operation and, where needed, explains both input and user prompts. Additionally, some of the programs have program notes explaining certain features which make the programs more flexible. Following the text is a sample run of the program itself and, finally, the listing. Most of the programs also have references which provide additional information on the subject matter.

We have written these programs to run on the Apple II microcomputer, but with a few modifications they will run on most of the popular microcomputers available today, including PET/CBM[2], Atari 400 and 800[3], and TRS-80[4].

None of the programs requires a mass storage device (disk or tape) for storing data. Thus the widely varying methods of accessing data files in BASIC[5] are not a problem. Of course, you will want to store the programs on disk or tape once you have them typed in. But this is a straightforward procedure adequately covered in your computer owner's manual.

How to Use These Programs

Follow the steps listed below to use any of the programs in this book.

1. Read the program write-up and familiarize yourself with how the program works. Read the references if they will give you a better understanding of the subject matter which the program addresses. Be sure the program does what you need it to do before going any further.

2. Type the program listing into your computer. Since the remark statements (those that begin with REM) are not essential to program operation, you need not type them in. This will save you time, and programs will take up less space (on most computers). The programs may even run marginally faster. But if you plan to modify a program extensively, you may be better off including its remarks, since they can be very helpful in tracing program logic flow during debugging.

3. Check your program listing carefully for errors. Compare it line-by-line and character-by-character with the published listing. Correct any discrepancies.

4. Save the program on tape or disk. Do it now, before you run the program. That way you can easily retrieve it in the event that something happens while you are running it.

5. Run the example exactly as shown in the sample run. If you have done everything right to this point, the results should be very similar to those published. Your answers will differ slightly from those in the book if your computer has a different level of numerical precision than ours.

6. If your answers differ markedly from ours, or your program does not run at all (i.e., you get some sort of error message), it is time for some detective work. First, double-check and triple-check your listing against the published one. We cannot overemphasize the importance of this scrutiny. Check for missing program lines and incorrect line numbers. Make sure you have

1. Apple II is a registered trademark of Apple Computer Inc.
2. The names PET and CBM are registered trademarks of Commodore Business Machines, Inc.
3. Atari 400 and Atari 800 are registered trademarks of Atari, Inc.
4. TRS-80 is a registered trademark of Tandy Corporation.
5. BASIC is a registered trademark of the Trustees of Dartmouth College.

entered the right letter or digit. It is easy to confuse zeros and O's, ones and I's, twos and Z's, fives and S's, and U's and V's.

If reviewing your listing doesn't disclose any typographical errors, check the appendix. See if any of the BASIC irregularities discussed there apply to your computer. If so, apply the suggested changes to your program and rerun it.

By now your programs should be running correctly. If not, have someone else look your program over. Often another set of eyes can see things that you will repeatedly miss. Try putting the program aside for a while and coming back to it. After a short break, errors you didn't see before may be obvious.

There is always the possibility that the program may still be incompatible with your BASIC. Detecting this requires some knowledge of programming. Compare the syntax of the program from the book with your BASIC syntax. Differences may be easy to spot or they may be subtle ones. Experience, ingenuity, and knowledge of your BASIC's quirks are your best tools at this point.

7. As a further test of your program, run the practice problems. Compare your answers with those in the book. Here again, they should be very close, though some slight discrepancy is not at all unusual.

Program Errors

If you encounter an error or program difficulty which you believe is not your fault, we would like to hear about it. Please write the publishers, and include the following information:

- A description of the error
- Data entered which caused the error
- A source listing of your program, from your computer (if possible)
- Any corrections you have

Contents

Appendix

1
Interpolation

Lagrangian Interpolation
Cubic Spline Interpolation

2

Lagrangian Interpolation

This program computes Y coordinates of points on a curve given their X coordinates. You must input coordinates of known points on the curve, no two having the same abscissa.

The computations are performed using the Lagrange method of interpolation.

The number of known points on the curve which may be entered in the program is limited to 50. You may increase or decrease this limit by altering line 30 as follows:

```
30 DIM X(P), Y(P)
```

where $P = $ the number of known points on a curve.

Examples

Consider the curve $Y = X^3 - 3X + 3$. You know that the points $(-3, -15)$, $(-2, 1)$, $(-1, 5)$, $(0, 3)$, $(1, 1)$, $(2, 5)$, and $(3, 21)$ are on the curve. What is the value of Y when $X = -1.65$ and 0.2?

```
LAGRANGIAN INTERPOLATION

NUMBER OF KNOWN POINTS?7
X,Y OF POINT1?-3,-15
X,Y OF POINT2?-2,1
X,Y OF POINT3?-1,5
X,Y OF POINT4?0,3
X,Y OF POINT5?1,1
X,Y OF POINT6?2,5
X,Y OF POINT7?3,21


INTERPOLATE: X= ?-1.65
             Y=   3.457875

MORE POINTS HERE(1=YES,0=NO)?1

INTERPOLATE: X= ?.2
             Y=   2.408
```

Given the following points from a sine curve, what is the sine of -2.47 and the sine of 1.5?

$(-5, 0.958)$	$(0, 0)$
$(-4, 0.757)$	$(1, 0.841)$
$(-3, -0.141)$	$(2, 0.909)$
$(-2, -0.909)$	$(3, 0.141)$
$(-1, -0.841)$	$(4, -0.757)$
	$(5, -0.959)$

```
MORE POINTS HERE(1=YES,0=NO)?0
ANOTHER CURVE?1
NUMBER OF KNOWN POINTS?11
X,Y OF POINT1?-5,.958
X,Y OF POINT2?-4,.757
X,Y OF POINT3?-3,-.141
```

```
X,Y OF POINT4?-2,-.909
X,Y OF POINT5?-1,-.841
X,Y OF POINT6?0,0
X,Y OF POINT7?1,.841
X,Y OF POINT8?2,.909
X,Y OF POINT9?3,.141
X,Y OF POINT10?4,-.757
X,Y OF POINT11?5,-.959


INTERPOLATE: X= ?-2.47
             Y=   -.621839596


MORE POINTS HERE(1=YES,0=NO)?1

INTERPOLATE: X= ?1.5
             Y=   .9971638


MORE POINTS HERE(1=YES,0=NO)?0
ANOTHER CURVE?0
```

Program Listing

```
5   HOME
10    PRINT "LAGRANGIAN INTERPOLATION"
20    PRINT
30    DIM X(50),Y(50)
40    PRINT "NUMBER OF KNOWN POINTS";
50    INPUT P
60    FOR I = 1 TO P
70    PRINT "X,Y OF POINT";I;
80    INPUT X(I),Y(I)
90    NEXT I
100   PRINT
110   PRINT "INTERPOLATE: X= ";
120   INPUT A
130 B = 0
140   FOR J = 1 TO P
150 T = 1
160   FOR I = 1 TO P
170   IF I = J THEN 190
180 T = T * (A - X(I)) / (X(J) - X(I))
190   NEXT I
200 B = B + T * Y(J)
210   NEXT J
220   PRINT "              Y=   ";B
230   PRINT
240   PRINT "MORE POINTS HERE";
245   PRINT "(1=YES,0=NO)";
250   INPUT C
260   IF C = 1 THEN 100
270   PRINT "ANOTHER CURVE";
280   INPUT C
290   IF C = 1 THEN 40
300   END
```

Cubic Spline Interpolation

This program computes either the value of Y computed for an increasing X entered from the keyboard, or a list of X, Y pairs computed for a fixed increment in X. The first mode is useful for reading values from a calibration curve, while the second would be commonly used in generating a list of points for plotting.

The cubic spline method uses a cubic (3rd degree) polynomial to interpolate between each pair of data points. A different polynomial is used for each interval, and each one is constrained to pass through the original data with the same slope as the data. In this program, the slopes are computed by finding the slope of the parabola that passes through each data point and its two nearest neighbors.

Because the program forces the computed values to equal the original data at the approximate X values, it is necessary to start with precise values, or else first smooth the data.

Examples

The average price for gasoline in January of each year from 1973 to 1980 is given in the table below.

Year	1973	1974	1975	1976	1977	1978	1979	1980
Price of Gas ($)	0.371	0.484	0.524	0.580	0.586	0.588	0.652	1.111

Estimate the price of gasoline in June 1975, September 1977, March 1979, and April 1974.

```
CUBIC SPLINE INTERPOLATION


ENTER NUMBER OF DATA POINTS: ?8

ENTER EACH DATA POINT AS X,Y

POINT # 1   X,Y = ?1973,.371
POINT # 2   X,Y = ?1974,.484
POINT # 3   X,Y = ?1975,.524
POINT # 4   X,Y = ?1976,.580
POINT # 5   X,Y = ?1977,.586
POINT # 6   X,Y = ?1978,.588
POINT # 7   X,Y = ?1979,.652
POINT # 8   X,Y = ?1980,1.111


  ENTER MODE #:
    1 = SINGLE POINT
    2 = INCREMENT  X
    3 = ENTER NEW DATA
    4 = STOP
WHICH? ?1
```

```
ENTER X VALUES ONE BY ONE.
RETURN TO MENU IF X IS OUTSIDE
OF RANGE X(1) TO X(N).


X = ?1975.5
     Y(X) = .554124998

X = ?1977.75
     Y(X) = .583234376

X = ?1979.25
     Y(X) = .729718729

X = ?1974.33
     Y(X) = .502023311


X = ?1
TOO SMALL


 ENTER MODE #:
   1 = SINGLE POINT
   2 = INCREMENT  X
   3 = ENTER NEW DATA
   4 = STOP
WHICH? ?2

ENTER X INCREMENT ?.25

1973              .371
1973.25           .406093752
1973.5            .436625002
1973.75           .462593751
1974              .484
1974.25           .498757812
1974.5            .5075625
1974.75           .514585937
1975              .524
1975.25           .538046874
1975.5            .554124998
1975.75           .569140622
1976              .58
1976.25           .585109378
1976.5            .586375003
1976.75           .585953126
1977              .586
1977.25           .585328125
1977.5            .583375001
1977.75           .583234376
1978              .588
1978.25           .590382817
1978.5            .591437514
1978.75           .606773454
1979              .652
```

```
1979.25          .729718729
1979.5           .832124973
1979.75          .95921873
```

Program Listing

```
100    HOME
210    PRINT "CUBIC SPLINE INTERPOLATION"
370    PRINT
380    PRINT
390    DIM X(100),Y(100),D(100),F(100),G(100),H(100)
400    PRINT "ENTER NUMBER OF DATA POINTS: ";
410    INPUT N
420    PRINT
430    PRINT "ENTER EACH DATA POINT AS X,Y"
440    PRINT
449    REM  ==INPUT DATA==
450    FOR R = 1 TO N
460    PRINT "POINT # ";R;"   X,Y = ";
470    INPUT X(R),Y(R)
480    IF R = 1 THEN 520
490    IF X(R) > X(R - 1) THEN 520
500    PRINT "NOT ALLOWED"
510    GOTO 460
520    NEXT R
529    REM ==COMPUTE DERIVATIVE==
530    GOSUB 1020
538    REM  ==COMPUTE CUBIC SPLINE COEFFICIENTS==
540    GOSUB 1180
550    PRINT
560    PRINT
570    PRINT " ENTER MODE #: "
580    PRINT "    1 = SINGLE POINT "
590    PRINT "    2 = INCREMENT  X"
600    PRINT "    3 = ENTER NEW DATA"
610    PRINT "    4 = STOP"
620    PRINT "WHICH? ";
630    INPUT Q
640    IF Q < 1 THEN 620
645    IF Q = 1 THEN 690
650    IF Q = 2 THEN 890
655    IF Q = 3 THEN 400
660    PRINT 670
670    PRINT "END OF PROGRAM"
680    GOTO 1280
687    REM === MODE 1: SINGLE X ===
690    PRINT
700    PRINT "ENTER X VALUES ONE BY ONE."
710    PRINT "RETURN TO MENU IF X IS OUTSIDE"
720    PRINT "OF RANGE X(1) TO X(N)."
730    PRINT
740    PRINT
750    PRINT "X = ";
760    INPUT J
```

```
770    IF J < X(1) THEN 870
779    REM ==FIND PROPER INTERVAL==
780    FOR K = 2 TO N
790    IF J < X(K) THEN 830
800    NEXT K
810    PRINT "TOO LARGE"
820    GOTO 880
830 K = K - 1
840    GOSUB 1250
850    PRINT "       Y(X) = ";V
860    GOTO 740
870    PRINT "TOO SMALL"
880    GOTO 550
886    REM ==MODE 2: INCREMENT X==
890    PRINT
900    PRINT "ENTER X INCREMENT ";
910    INPUT U
920    PRINT
930 J = X(1)
940 K = 1
948    REM ==COMPUTE INTERPOLATED==
949    REM        === VALUE ===
950    GOSUB 1250
951    GOSUB 1250
958    REM =PRINT COORDINATES OF INTERPOLATED POINT=
959    REM   ==== FOR PLOTTING REPLACE J,V WITH AN ARRAY ====
960    PRINT J,V
970 J = J + U
980    IF J < X(K + 1) THEN 950
990 K = K + 1
1000    IF K = N THEN 550
1010    GOTO 950
1016    REM ====COMPUTE SLOPES====
1020 J = 2
1030    GOSUB 1120
1040 D(1) = 2 * A * X(1) + B
1050 D(2) = 2 * A * X(2) + B
1060    FOR J = 3 TO N - 1
1070    GOSUB 1120
1080 D(J) = 2 * A * X(J) + B
1090    NEXT J
1100 D(N) = 2 * A * X(N) + B
1110    RETURN
1116    REM ===COMPUTE COEFFICIENTS OF PARABOLA===
1120 A = (Y(J - 1) - Y(J)) / (X(J - 1) - X(J))
1130 A = A - (Y(J) - Y(J + 1)) / (X(J) - X(J + 1))
1140 A = A / (X(J - 1) - X(J + 1))
1150 B = (Y(J - 1) - Y(J)) / (X(J - 1) - X(J))
1160 B = B - A * (X(J - 1) + X(J))
1170    RETURN
1176    REM ==CUBIC SPLINE COEFFICIENTS==
1180    FOR J = 1 TO N - 1
1190 F(J) = D(J) * (X(J + 1) - X(J))
1200 G(J) = 3 * Y(J + 1) - D(J + 1) * (X(J + 1) - X(J))
1210 G(J) = G(J) - 3 * Y(J) - 2 * F(J)
```

```
1220 H(J) = Y(J + 1) - Y(J) - F(J) - G(J)
1230  NEXT J
1246  REM ==EVALUATE POLYNOMIAL==
1250 W = (J - X(K)) / (X(K + 1) - X(K))
1260 V = Y(K) + F(K) * W + G(K) * W * W + H(K) * W * W * W
1270  RETURN
1280  END
```

References

Akima, H. "A new method of interpolation and smooth curve fitting based on local procedures."
   Journal of the Association for Computing Machinery 17 (1970), pp. 589-602.

Monro, D. M. "Interpolation methods for surface mapping." *Computer Programs in Biomedicine* 11
   (1980), 145-57.

2
Regression

Linear Regression

This program fits a straight line to a given set of coordinates using the method of least squares. The equation of the line, coefficient of determination, coefficient of correlation, and standard error of estimate are printed. Once the line has been fitted you may predict values of Y for given values of X.

Example

The table below shows the height and weight of 11 male college students. Fit a curve to these points. How much would the average 70'' and 72'' male student weigh?

Height (in.)	71	73	64	65	61	70	65	72	63	67	64
Weight (lbs.)	160	183	154	168	159	180	145	210	132	168	141

```
RUN
LINEAR REGRESSION

NUMBER OF KNOWN POINTS11
X,Y OF POINT1?71,160
X,Y OF POINT2?73,183
X,Y OF POINT3?64,154
X,Y OF POINT4?65,168
X,Y OF POINT5?61,159
X,Y OF POINT6?70,180
X,Y OF POINT7?65,145
X,Y OF POINT8?72,210
X,Y OF POINT9?63,132
X,Y OF POINT10?67,168
X,Y OF POINT11?64,141

F(X) =-106.791727+ (4.04722312* X )

COEFFICIENT OF DETERMINATION (R^2)=.556260313
COEFFICIENT OF CORRELATION=.745828608
STANDARD ERROR OF ESTIMATE =15.4134854

INTERPOLATION: (ENTER X=0 TO END PROGRAM)
X = 70
Y = 176.513892

X = 72
Y = 184.608338

X = 0
```

Program Listing

```
5   HOME
10   PRINT "LINEAR REGRESSION"
```

```
20    PRINT
30    INPUT "NUMBER OF KNOWN POINTS";N
40  J = 0:K = 0:L = 0:M = 0:R2 = 0
99    REM ==LOOP TO ENTER COORDINATES OF POINTS==
100   FOR T = 1 TO N
110    PRINT "X,Y OF POINT";T;
120    INPUT X,Y
129    REM ==ACCUMULATE INTERMEDIATE SUMS==
130  J = J + X
140  K = K + Y
150  L = L + X ^ 2
160  M = M + Y ^ 2
170  R2 = R2 + X * Y
180    NEXT T
189    REM ==COMPUTE CURVE COEFFICIENT==
190  B = (N * R2 - K * J) / (N * L - J ^ 2)
200  A = (K - B * J) / N
210    PRINT
220    PRINT "F(X) =";A;"+ (";B;"* X )"
229    REM ==COMPUTE REGRESSION ANALYSIS==
230  J = B * (R2 - J * K / N)
240  M = M - K ^ 2 / N
250  K = M - J
260    PRINT
270  R2 = J / M
280    PRINT "COEFFICIENT OF DETERMINATION (R^2)=";R2
290    PRINT "COEFFICIENT OF CORRELATION="; SQR (R2)
300    PRINT "STANDARD ERROR OF ESTIMATE ="; SQR (K / (N - 2))
310    PRINT
319    REM ==ESTIMATE Y-COORDINATES OF POINTS WITH ENTERED X-COORDINATES==
320    PRINT "INTERPOLATION: (ENTER X=0 TO END PROGRAM)"
330    INPUT "X = ";X
349    REM ==RESTART OR END PROGRAM==
350    IF X = 0 THEN 390
360    PRINT "Y = ";A + B * X
370    PRINT
380    GOTO 330
390    END
```

Geometric Regression

This program fits a geometric curve to a set of coordinates using the method of least squares. The equation, coefficient of determination, coefficient of correlation, and standard error of estimate are printed.

You must provide the X and Y coordinates of known data points. Once the curve has been fitted you may predict values of Y for given values of X.

Example

The table below gives the pressures of a gas measured at various volumes in an experiment. The relationship between pressure and volume of a gas is expressed by the following formula:

$$PV^K = C$$

where: P = pressure

V = volume

C and K are constants.

This formula can be rewritten in standard geometric form:

$$P = CV^{-K}$$

Note that the exponent is negative, which accounts for the negative exponents the program calculates.

Fit a geometric curve to the data and estimate the pressure of 90 cubic inches of the gas.

Volume	56.1	60.7	73.2	88.3	120.1	187.5
Pressure	57.0	51.0	39.2	30.2	19.6	10.5

```
RUN
GEOMETRIC REGRESSION

NUMBER OF KNOWN POINTS?6
X,Y OF POINT1?56.1,57.0
X,Y OF POINT2?60.7,51.0
X,Y OF POINT3?73.2,39.2
X,Y OF POINT4?88.3,30.2
X,Y OF POINT5?120.1,19.6
X,Y OF POINT6?187.5,10.5

F(X) =16103.7139* X^-1.40155091

COEFFICIENT OF DETERMINATION
(R^2) =.999999206

COEFFICIENT OF CORRELATION =.999999603

STANDARD ERROR OF ESTIMATE = 6.37591016E-04
```

```
INTERPOLATION: (X=0 TO END)
X =?90
Y =29.3734983
```

Program Listing

```
5    HOME
10    PRINT "GEOMETRIC REGRESSION"
20    PRINT
30    PRINT "NUMBER OF KNOWN POINTS";
40    INPUT N
50 J = 0
60 K = 0
70 L = 0
80 M = 0
90 R2 = 0
100   FOR I = 1 TO N
110   PRINT "X,Y OF POINT";I;
120    INPUT X,Y
130 Y =   LOG (Y)
140 X =   LOG (X)
150 J = J + X
160 K = K + Y
170 L = L + X ^ 2
180 M = M + Y ^ 2
190 R2 = R2 + X * Y
200    NEXT I
210 B = (N * R2 - K * J) / (N * L - J ^ 2)
220 A = (K - B * J) / N
230    PRINT
240    PRINT "F(X) ="; EXP (A);"* X^";B
250 J = B * (R2 - J * K / N)
260 M = M - K ^ 2 / N
270 K = M - J
280    PRINT
290 R2 = J / M
300    PRINT "COEFFICIENT OF ";
301    PRINT "DETERMINATION"
305    PRINT "(R^2) =";R2
306    PRINT
310    PRINT "COEFFICIENT OF ";
311    PRINT "CORRELATION ="; SQR (R2)
315    PRINT
320    PRINT "STANDARD ERROR OF ";
325    PRINT "ESTIMATE = "; SQR (K / (N - 2))
326    PRINT
330    PRINT
340    PRINT "INTERPOLATION: ";
341    PRINT "(X=0 TO END)"
350    PRINT "X =";
360    INPUT X
370    IF X = 0 THEN 410
380    PRINT "Y ="; EXP (A) * X ^ B
390    PRINT
400    GOTO 350
410    END
```

Exponential Regression

This program finds the coefficients of an equation for an exponential curve. The equation is in the following form:

$$f(X) = Ae^{BX}$$

where A and B are the calculated coefficients.

The equation coefficients, coefficient of determination, coefficient of correlation, and standard error of estimate are printed.

You must provide the X and Y coordinates for known data points. Once the curve has been fitted you may predict values of Y for given values of X.

Example

The table below shows the number of bacteria present in a culture at various points in time. Fit an exponential curve to the data and estimate the number of bacteria after 7 hours.

Number of hours	0	1	2	3	4	5	6
Number of bacteria	25	38	58	89	135	206	315

```
RUN
EXPONENTIAL REGRESSION

NUMBER OF KNOWN POINTS?7
X,Y OF POINT1?0,25
X,Y OF POINT2?1,38
X,Y OF POINT3?2,58
X,Y OF POINT4?3,89
X,Y OF POINT5?4,135
X,Y OF POINT6?5,206
X,Y OF POINT7?6,315

A =24.9616634
B =.422375081

COEFFICIENT OF DETERMINATION
(R^2) =.999993572

COEFFICIENT OF CORRELATION =.999996786

STANDARD ERROR OF ESTIMATE =2.53424554E-03


INTERPOLATION: (X=0 TO END)
X =?7
Y =480.086716

X =?0
```

Program Listing

```
5    HOME
10   PRINT "EXPONENTIAL REGRESSION"
20   PRINT
30   PRINT "NUMBER OF KNOWN POINTS";
40   INPUT N
50 J = 0
60 K = 0
70 L = 0
80 M = 0
90 R2 = 0
99   REM ==ENTER COORDINATES OF DATA POINTS==
100   FOR I = 1 TO N
110   PRINT "X,Y OF POINT";I;
120   INPUT X,Y
129   REM ==ACCUMULATE INTERMEDIATE VALUES==
130 Y =   LOG (Y)
140 J = J + X
150 K = K + Y
160 L = L + X ^ 2
170 M = M + Y ^ 2
180 R2 = R2 + X * Y
190   NEXT I
199   REM ==CALCULATE INTERMEDIATE VALUES==
200 B = (N * R2 - K * J) / (N * L - J ^ 2)
210 A = (K - B * J) / N
220   PRINT
230   PRINT "A ="; EXP (A)
240   PRINT "B =";B
249   REM ==CALCULATE REGRESSION TABLE VALUES==
250 J = B * (R2 - J * K / N)
260 M = M - K ^ 2 / N
270 K = M - J
280   PRINT
290 R2 = J / M
300   PRINT "COEFFICIENT OF ";
301   PRINT "DETERMINATION"
305   PRINT "(R^2) =";R2
306   PRINT
310   PRINT "COEFFICIENT OF ";
311   PRINT "CORRELATION =";
315   PRINT  SQR (R2)
316   PRINT
320   PRINT "STANDARD ERROR OF ";
321   PRINT "ESTIMATE ="; SQR (K / (N - 2))
326   PRINT
330   PRINT
340   PRINT "INTERPOLATION: ";
341   PRINT "(X=0 TO END)"
350   PRINT "X =";
360   INPUT X
370   IF X = 0 THEN 410
380   PRINT "Y ="; EXP (A) *  EXP (B * X)
390   PRINT
400   GOTO 350
410   END
```

*N*th Order Regression

This program finds the coefficients of an *N*th order equation using the method of least squares. The equation is of the following form:

$$Y = C + A_1 X + A_2 X^2 + \ldots + A_n X^n$$

where: Y = dependent variable
C = constant
$A_1, A_2, \ldots, A_n$ = coefficients of independent variables $X, X^2 \ldots, X^n$, respectively

The equation coefficients, coefficient of determination, coefficient of correlation, and standard error of estimate are printed.

You must provide the X and Y coordinates for known data points. Once the equation has been computed you may predict values of Y for given values of X.

The dimension statement at line 20 limits the degree of the equation. You can change this limit according to the following scheme:

```
20   DIM A(2×D + 1),R(D + 1,D + 2),T(D + 2)
```

where D = maximum degree of equation.

Example

The table below gives the stopping distance (reaction plus braking distance) of an automobile at various speeds. Fit an exponential curve to the data. Estimate the stopping distance at 55 m.p.h.

M.p.h.	20	30	40	50	60	70
Stopping distance	54	90	138	206	292	396

```
RUN
N'TH ORDER REGRESSION

DEGREE OF EQUATION?2
NUMBER OF KNOWN POINTS?6
X,Y OF POINT1?20,54
X,Y OF POINT2?30,90
X,Y OF POINT3?40,138
X,Y OF POINT4?50,206
X,Y OF POINT5?60,292
X,Y OF POINT6?70,396

        CONSTANT =41.7714472
1DEGREE COEFFICIENT =-1.09571524
2DEGREE COEFFICIENT =.0878571531


COEFFICIENT OF DETERMINATION
(R^2) =.999927959
CORRELATION COEFFICIENT =.999963979
```

STANDARD ERROR ESTIMATE =1.42094106

INTERPOLATION: (ENTER 0 TO END)
X =?55
Y =247.274998

X =?0

Program Listing

```
5    HOME
10   PRINT "N'TH ORDER REGRESSION"
20   DIM A(13),R(7,8),T(8)
30   PRINT
40   PRINT "DEGREE OF EQUATION";
50   INPUT D
60   PRINT "NUMBER OF KNOWN POINTS";
70   INPUT N
80  A(1) = N
90   FOR I = 1 TO N
100   PRINT "X,Y OF POINT";I;
110   INPUT X,Y
120   FOR J = 2 TO 2 * D + 1
130  A(J) = A(J) + X ^ (J - 1)
140   NEXT J
150   FOR K = 1 TO D + 1
160  R(K,D + 2) = T(K) + Y * X ^ (K - 1)
170  T(K) = T(K) + Y * X ^ (K - 1)
180   NEXT K
190  T(D + 2) = T(D + 2) + Y ^ 2
200   NEXT I
210   FOR J = 1 TO D + 1
220   FOR K = 1 TO D + 1
230  R(J,K) = A(J + K - 1)
240   NEXT K
250   NEXT J
260   FOR J = 1 TO D + 1
270   FOR K = J TO D + 1
280   IF R(K,J) < > 0 THEN 320
290   NEXT K
300   PRINT "NO UNIQUE SOLUTION"
310   GOTO 790
320   FOR I = 1 TO D + 2
330  S = R(J,I)
340  R(J,I) = R(K,I)
350  R(K,I) = S
360   NEXT I
370  Z = 1 / R(J,J)
380   FOR I = 1 TO D + 2
390  R(J,I) = Z * R(J,I)
400   NEXT I
410   FOR K = 1 TO D + 1
420   IF K = J THEN 470
430  Z =  - R(K,J)
440   FOR I = 1 TO D + 2
450  R(K,I) = R(K,I) + Z * R(J,I)
```

```
460    NEXT I
470    NEXT K
480    NEXT J
490    PRINT
495    PRINT "                    CONSTANT =";
496    PRINT R(1,D + 2)
500    FOR J = 1 TO D
510    PRINT J;"DEGREE COEFFICIENT =";
511    PRINT R(J + 1,D + 2)
520    NEXT J
530    PRINT
540 P = 0
550    FOR J = 2 TO D + 1
560 P = P + R(J,D + 2) * (T(J) - A(J) * T(1) / N)
570    NEXT J
580 Q = T(D + 2) - T(1) ^ 2 / N
590 Z = Q - P
600 I = N - D - 1
620    PRINT
630 J = P / Q
640    PRINT "COEFFICIENT OF ";
641    PRINT "DETERMINATION"
645    PRINT "(R^2) =";J
650    PRINT "CORRELATION COEFFICIENT =";
651    PRINT  SQR (J)
660    PRINT "STANDARD ERROR ESTIMATE =";
661    PRINT  SQR (Z / I)
670    PRINT
680    PRINT "INTERPOLATION: ";
681    PRINT "(ENTER 0 TO END)"
690 P = R(1,D + 2)
700    PRINT "X =";
710    INPUT X
720    IF X = 0 THEN 790
730    FOR J = 1 TO D
740 P = P + R(J + 1,D + 2) * X ^ J
750    NEXT J
760    PRINT "Y =";P
770    PRINT
780    GOTO 690
790    END
```

Regression of an Arbitrary Function in One Variable

This program fits a user-defined function to a set of data points using an iteration technique. The programmer supplies the data points, the general form of the equation, estimates of the equation coefficients, and the desired root mean square error. The program refines the coefficients until the error reaches the desired value. . .or until the user loses patience.

Root mean square error is also known as standard deviation, and is given by the formula:

$$R_1 = \sqrt{\frac{\sum_{J=1}^{N} (F_J - Y_J)^2}{N}}$$

where R_1 = Root mean square error or standard deviation
N = number of data points
Y_J = input values of Y
F_J = calculated values of Y

Note: This program will use large amounts of time running if:

1. The number of data points is large,
2. The number of coefficients is greater than two, or
3. The starting value of the coefficients is poorly selected.

Example

Given a set of data points (X_n, Y_n) that are known to fit a straight line:

$X_1 = 1$	$Y_1 = 7$
$X_2 = 2$	$Y_2 = 11$
$X_3 = 3$	$Y_3 = 15$
$X_4 = 4$	$Y_4 = 19$
$X_5 = 5$	$Y_5 = 23$

Using the formula for a straight line:

$$Y = A_1 + A_2 \times X$$

Calculate A_1 and A_2 (the unknown coefficients).

Before running the program it will be necessary to input the general form of the equation at line 1100 in the form:

$$1100\ Y = A_1 + A_2 \times X$$

When the program is run, the user will be asked for the number of data points N, the number of equation coefficients M, the desired root mean square error R, and estimates of the equation coefficients: $A_1, A_2, A_3, . . ., A_n$.

```
RUN
REGRESSION OF AN ARBITRARY FUNCTION
IN ONE VARIABLE
```

```
NUMBER OF DATA POINTS?10

NUMBER OF COEFFICIENTS?3

DESIRED MAX ERROR?.02

ENTER X,Y
?.1,.993
INPUT NEXT X,Y
?.2,1.31
INPUT NEXT X,Y
?.3,1.62
INPUT NEXT X,Y
?.4,1.92
INPUT NEXT X,Y
?.5,2.19
INPUT NEXT X,Y
?.6,2.45
INPUT NEXT X,Y
?.7,2.68
INPUT NEXT X,Y
?.8,2.88
INPUT NEXT X,Y
?.9,3.05
INPUT NEXT X,Y
?1.0,3.19


UNKNOWN COEFFICIENTS:


ESTIMATE   A1= ?.7
ESTIMATE   A2= ?3
ESTIMATE   A3= ?1.2


            -RUNNING-


COEFFICIENT A1=.610071909
COEFFICIENT A2=2.70813823
COEFFICIENT A3=1.25265058

ROOT MEAN SQUARE ERROR = .019146606

INTERPOLATE (1=YES/0=NO) ?1

X = ?.7
Y = 2.69190305

INTERPOLATE (1=YES/0=NO) ?1

X = ?.334
Y = 1.71034935

INTERPOLATE (1=YES/0=NO) ?0
```

Program Listing

```
5    HOME
100   PRINT "REGRESSION OF AN ARBITRARY FUNCTION"
105   PRINT "IN ONE VARIABLE"
110   PRINT
120   REM ==ENTER EQUATION AT LINE 1100==
130   DIM A(5),D(5),X(50),Y(50)
140   PRINT "NUMBER OF DATA POINTS";
150   INPUT N
160   PRINT
170   PRINT "NUMBER OF COEFFICIENTS";
180   INPUT M
190   PRINT
200   PRINT "DESIRED MAX ERROR";
210   INPUT R
220   PRINT
230   REM ==ENTER DATA POINTS==
235   PRINT "ENTER X,Y"
237   INPUT X(1),Y(1)
240   FOR J = 2 TO N
245   PRINT "INPUT NEXT X,Y"
250   INPUT X(J),Y(J)
260   NEXT J
270   REM ==INPUT COEFFICIENT ESTIMATES==
275   PRINT : PRINT
277   PRINT "UNKNOWN COEFFICIENTS:"
279   PRINT : PRINT
280   FOR K = 1 TO M
290   PRINT "ESTIMATE   A";K;"= ";
300   INPUT A(K)
310 D(K) = A(K) / 10
320   NEXT K
330   REM ==START ITERATION==
340   FOR K = 1 TO M
350   GOSUB 1000
360 R1 = R8
370 N7 = 0
380 N8 = 0
390 D = D(K)
400 A(K) = A(K) + D
410   GOSUB 1000
420 R2 = R8
430   IF R1 > R2 THEN 490
440 D =   - D
450 A(K) = A(K) + D
460 R9 = R1
470 R1 = R2
480 R2 = R9
490 N7 = N7 + 1
500 N8 = N8 + 1
510 A(K) = A(K) + D
520   GOSUB 1000
530 R3 = R8
540   PRINT
550   PRINT "R1="R1
```

```
560    PRINT "R2="R2
570    PRINT "R3="R3
580    PRINT
590    FOR L = 1 TO M
600    PRINT "A"L" = "A(L)
610    NEXT L
620    PRINT
630    IF R3 > R2 THEN 700
640 R1 = R2
650 R2 = R3
660    IF R1 = R2 AND R2 = R3 THEN 750
670    IF N8 < 10 THEN 490
680 D(K) = D(K) * 10
690    GOTO 350
700 A(K) = A(K) - D * (1 / (1 + (R1 - R2) / (R3 - R2)) + .5)
710 D(K) = D(K) * N7 / 5
720    NEXT K
730    IF R8 < R THEN 750
740    GOTO 340
750    FOR K = 1 TO M
760    PRINT "COEFFICIENT A";K;"=";A(K)
770    NEXT K
780    PRINT
790    PRINT "ROOT MEAN SQUARE ERROR = ";R8
800    PRINT
810    PRINT "INTERPOLATE (1=YES/0=NO) ";
820    INPUT Z
830    IF Z = 0 THEN 1300
840    PRINT
850    PRINT "X = ";
860    INPUT X(1)
870 N = 1
880    GOSUB 1000
890    PRINT "Y = "Y
900    GOTO 800
910    REM ==PRINT THE RESULTS==
1001   REM ==SUBROUTINE FOR STANDARD DEVIATION==
1010 R8 = 0
1020   FOR J = 1 TO N
1030 A1 = A(1)
1040 A2 = A(2)
1050 A3 = A(3)
1060 A4 = A(4)
1070 A5 = A(5)
1080 X = X(J)
1090   REM ==ENTER EQUATION AT 1100==
1100 Y = A1 + A2 * X
1110 R9 = Y - Y(J)
1120 R8 = R8 + R9 ^ 2
1130   NEXT J
1140 R8 =   SQR (R8 / N)
1150   RETURN
1199   REM ==ENTER DATA POINTS BELOW==
1300   END
```

Polynomial Regression in Two Variables

This program generates a polynomial in two variables using a set of data points utilizing the method of least squares. The solution provides the coefficients A_1 to A_9 of the equation:

$$D = A_1 + A_2 X + A_3 X^2 + A_4 Y + A_5 Y^2 + A_6 XY + A_7 X^2 Y + A_8 XY^2 + A_9 X^2 Y^2$$

where:

$$D = \text{dependent variable}$$
$$X, Y = \text{independent variables}$$
$$A_1, A_2, \ldots, A_9 = \text{coefficients of independent variables}$$

You must provide input data at lines 0-99 in the form:

$$1 \quad \text{DATA} \quad X_1, Y_1, D_1$$
$$2 \quad \text{DATA} \quad X_2, Y_2, D_2$$
$$\cdot$$
$$\cdot$$
$$\cdot$$
$$N \quad \text{DATA} \quad X_n, Y_n, D_n 11$$

The dimension statement at line 190 limits the number of data points the equation may contain to 50. You can change this limit according to the following scheme:

```
190 DIM X(N), Y(N), D(N)
```

where N = number of data points.

Note: This program will take several minutes to run if the number of input points is large.

The output of this program will print the nine equation coefficients followed by the standard deviation of the estimate. You may then predict values of D for any given values of X and Y.

The standard deviation calculation (also known as the root mean square error) is given by the equation:

$$R_2 = \sqrt{\frac{\sum\limits_{J=1}^{N} (DC_J - D_J)^2}{N}}$$

where:

R_2 = standard deviation
DC = calculated value of D
D = input value of D
N = number of data points

Example

The table below lists the performance of a centrifugal fan where the speed (revolutions per minute, RPM) is a function of capacity (cubic feet per minute, CFM) and static pressure (inches of water, SP). Capacity (CFM) and static pressure (SP) are the independent variables, and revolutions per minute (RPM) is the dependent variable.

Capacity	Static Pressure			
(CFM)	0.8	1.2	1.6	2.0
	Fan Speed (RPM)			
1400	873	1076	1253	1409
1600	877	1072	1245	1399
1800	888	1070	1239	1391
2000	904	1076	1235	1386

Data input is as follows:

```
 1  DATA  0.8, 1400, 873
 2  DATA  0.8, 1600, 877
              .
              .
              .
16  DATA  2.0, 2000, 1386
```

Find the coefficients of the polynomial and predict the speed that the fan must turn to deliver 1550 CFM at 1.76 SP.

```
RUN
POLYNOMIAL REGRESSION IN TWO VARIABLES

WHERE X AND Y ARE INDEPENDENT VARIABLES
AND D IS THE DEPENDENT VARIABLE

ENTER NUMBER OF DATA POINTS:
?16
INPUT X,Y,D
?0.8,1400,873
INPUT NEXT X,Y,D
?0.8,1600,877
INPUT NEXT X,Y,D
?0.8,1800,888
INPUT NEXT X,Y,D
?0.8,2000,904
INPUT NEXT X,Y,D
?1.2,1400,1076
INPUT NEXT X,Y,D
?1.2,1600,1072
INPUT NEXT X,Y,D
?1.2,1800,1070
INPUT NEXT X,Y,D
?1.2,2000,1076
INPUT NEXT X,Y,D
?1.6,1400,1253
INPUT NEXT X,Y,D
?1.6,1600,1245
INPUT NEXT X,Y,D
?1.6,1800,1239
INPUT NEXT X,Y,D
?1.6,2000,1235
INPUT NEXT X,Y,D
?2.0,1400,1409
```

```
INPUT NEXT X,Y,D
?2.0,1600,1399
INPUT NEXT X,Y,D
?2.0,1800,1391
INPUT NEXT X,Y,D
?2.0,2000,1386

          -RUNNING-


COEFFICIENTS A1 - A9 ARE :

530.120398
684.298471
-89.6199925
-.316912398
1.58849709E-04
.151420737
-.0296921539
-1.23892831E-04
2.91835925E-05

STANDARD DEVIATION = .54184341

X = 1.76
Y = 1550

D = 1311.13184

CONTINUE = 1    END = 0 ?0
```

Program Listing

```
10    HOME
100   PRINT "POLYNOMIAL REGRESSION IN TWO VARIABLES"
110   PRINT
160   PRINT "WHERE X AND Y ARE INDEPENDENT VARIABLES"
170   PRINT "AND D IS THE DEPENDENT VARIABLE"
180   PRINT
190   DIM X(50),Y(50),D(50)
200   DIM Z(9),A(9,10)
210   PRINT "ENTER NUMBER OF DATA POINTS:"
220   INPUT N
230   REM ==LOAD DATA POINTS==
233   PRINT "INPUT X,Y,D"
237   INPUT X(1),Y(1),D(1)
240   FOR J = 2 TO N
245   PRINT "INPUT NEXT X,Y,D"
250   INPUT X(J),Y(J),D(J)
260   NEXT J
270   REM ==ZERO-Z COUNTERS==
280   FOR K = 1 TO 9
290 Z(K) = 0
300   NEXT K
310   REM ==ZERO MATRIX COUNTERS==
```

```
320   FOR K = 1 TO 9
330   FOR L = 1 TO 10
340 A(K,L) = 0
350   NEXT L
360   NEXT K
370   REM ==LOAD MATRIX==
375   PRINT
380   PRINT "            -RUNNING-"
390   PRINT
400   FOR J = 1 TO N
420   GOSUB 2000
510   FOR K = 1 TO 9
520   FOR L = 1 TO 9
530 A(K,L) = A(K,L) + Z(K) * Z(L)
540   NEXT L
550 A(K,10) = A(K,10) + D(J) * Z(K)
560   NEXT K
570   NEXT J
580   REM ==SOLVE FOR COEFFICIENTS==
590   FOR K = 1 TO 9
610 P = A(K,K)
620 A(K,K) = 1
630   IF P = 0 THEN 800
640   FOR L = K + 1 TO 10
650 A(K,L) = A(K,L) / P
660   NEXT L
670 M = 1
680   IF M = K THEN 740
690 R = A(M,K)
700   FOR L = 1 TO 10
710 A(M,L) = A(M,L) - R * A(K,L)
720   NEXT L
730 A(M,K) = 0
740 M = M + 1
750   IF M = 10 THEN 770
760   GOTO 680
770   NEXT K
780   PRINT
790   GOTO 830
800   REM ==ERROR COMMENTS==
810   PRINT "ZERO DIAGONAL - NO SOLUTION "
820   GOTO 2030
830   REM ==PRINT COEFFICIENTS==
840   PRINT "COEFFICIENTS A1 - A9 ARE :"
850   PRINT
860   FOR K = 1 TO 9
870   PRINT A(K,10)
880   NEXT K
890   PRINT
900   REM ==CALCULATE STANDARD DEVIATION==
910 Q = 0
920   FOR J = 1 TO N
930   GOSUB 2000
935 DC = 0
940   FOR K = 1 TO 9
```

```
950 DC = DC + A(K,10) * Z(K)
960   NEXT K
970 Q = Q + (DC - D(J)) ^ 2
980   NEXT J
990 R =  SQR (Q / N)
995   PRINT "STANDARD DEVIATION = "R
998   PRINT
1000   REM ==INTERPOLATE==
1005 D = 0
1010   INPUT "X = ";X(50)
1020   INPUT "Y = ";Y(50)
1030   PRINT
1040 J = 50
1050   GOSUB 2000
1060   FOR K = 1 TO 9
1070 D = D + A(K,10) * Z(K)
1080   NEXT K
1090   PRINT "D = "D
1100   PRINT
1110   PRINT "CONTINUE = 1    END = 0 ";
1115   INPUT OK
1120   IF OK = 1 THEN 1000
1130   END
2000   REM ==SUBROUTINE==
2011 Z(1) = 1
2012 Z(2) = X(J)
2013 Z(3) = X(J) * X(J)
2014 Z(4) = Y(J)
2015 Z(5) = Y(J) * Y(J)
2016 Z(6) = X(J) * Y(J)
2017 Z(7) = Z(3) * Y(J)
2018 Z(8) = X(J) * Z(5)
2019 Z(9) = Z(3) * Z(5)
2020   RETURN
2030   END
```

Multiple Linear Regression

This program finds the coefficients of a multiple-variable linear equation using the method of least squares. The equation is of the following form:

$$Y = C + A_1 X_1 + A_2 X_2 + \ldots + A_n X_n$$

where: Y = dependent variable

C = constant

$A_1, A_2, \ldots, A_n$ = coefficients of independent variables $X_1, X_2, \ldots, X_n$

The constant and the coefficients are printed.

You must provide the X and Y coordinates of known data points. Once the equation has been found using the data you enter, you may predict values of the dependent variables for given values of the independent variables.

The dimension statement at line 30 limits the number of known data points the equation may contain. You can change this limit according to the following scheme:

```
30 DIM X(N + 1), S(N + 1), T(N + 1), A(N + 1, N + 2)
```

where N = the number of known data points.

Example

The table below shows the age, height, and weight of eight boys. Using weight as the dependent variable, fit a curve to the data. Estimate the weight of a seven-year-old boy who is 51 inches tall.

Age	8	9	6	10	8	9	9	7
Height	48	49	44	59	55	51	55	50
Weight	59	55	50	80	61	75	67	58

```
RUN
MULTIPLE LINEAR REGRESSION

NUMBER OF KNOWN POINTS?8
NUMBER OF INDEPENDENT VARIABLES?2
POINT1
 VARIABLE1?8
 VARIABLE2?48
 DEPENDENT VARIABLE?59
POINT2
 VARIABLE1?9
 VARIABLE2?49
 DEPENDENT VARIABLE?55
POINT3
 VARIABLE1?6
 VARIABLE2?44
 DEPENDENT VARIABLE?50
```

```
POINT4
  VARIABLE1?10
  VARIABLE2?59
  DEPENDENT VARIABLE?80
POINT5
  VARIABLE1?8
  VARIABLE2?55
  DEPENDENT VARIABLE?61
POINT6
  VARIABLE1?9
  VARIABLE2?51
  DEPENDENT VARIABLE?75
POINT7
  VARIABLE1?9
  VARIABLE2?55
  DEPENDENT VARIABLE?67
POINT8
  VARIABLE1?7
  VARIABLE2?50
  DEPENDENT VARIABLE?58

EQUATION COEFFICIENTS:
      CONSTANT:-15.7021277
VARIABLE(1):3.68085106
VARIABLE(2):.943262412

COEFFICIENT OF DETERMINATION
          (R^2) =.715697404
COEFFICIENT OF MULTIPLE
CORRELATION =.845989009
STANDARD ERROR OF ESTIMATE = 6.42887917

INTERPOLATION: (ENTER 0 TO END PROGRAM)
VARIABLE1?7
VARIABLE2?51
DEPENDENT VARIABLE =58.1702128

VARIABLE1?0


END OF PROGRAM
```

Program Listing

```
5    HOME
10   PRINT "MULTIPLE LINEAR REGRESSION"
20   PRINT
30   DIM X(9),S(9),T(9),A(9,10)
40   PRINT "NUMBER OF KNOWN POINTS";
50   INPUT N
60   PRINT "NUMBER OF INDEPENDENT VARIABLES";
70   INPUT V
80 X(1) = 1
90   FOR I = 1 TO N
100  PRINT "POINT";I
110  FOR J = 1 TO V
```

```
120    PRINT " VARIABLE";J;
130    INPUT X(J + 1)
131 M =   INT (T)
140    NEXT J
150    PRINT " DEPENDENT VARIABLE";
160    INPUT X(V + 2)
170    FOR K = 1 TO V + 1
180    FOR L = 1 TO V + 2
190 A(K,L) = A(K,L) + X(K) * X(L)
200 S(K) = A(K,V + 2)
210    NEXT L
220    NEXT K
230 S(V + 2) = S(V + 2) + X(V + 2) ^ 2
240    NEXT I
250    FOR I = 2 TO V + 1
260 T(I) = A(1,I)
270    NEXT I
280    FOR I = 1 TO V + 1
290 J = I
300    IF A(J,I) <  > 0 THEN 340
305 J = J + 1
310    IF J <  = V + 1 THEN 300
320    PRINT "NO UNIQUE SOLUTION"
330    GOTO 810
340    FOR K = 1 TO V + 2
350 B = A(I,K)
360 A(I,K) = A(J,K)
370 A(J,K) = B
380    NEXT K
390 Z = 1 / A(I,I)
400    FOR K = 1 TO V + 2
410 A(I,K) = Z * A(I,K)
420    NEXT K
430    FOR J = 1 TO V + 1
440    IF J = I THEN 490
450 Z =   - A(J,I)
460    FOR K = 1 TO V + 2
470 A(J,K) = A(J,K) + Z * A(I,K)
480    NEXT K
490    NEXT J
500    NEXT I
510    PRINT
520    PRINT "EQUATION COEFFICIENTS:"
525    PRINT "      CONSTANT:"A(1,V + 2)
530    FOR I = 2 TO V + 1
540    PRINT "VARIABLE(";I - 1;"):";A(I,V + 2)
550    NEXT I
560 P = 0
570    FOR I = 2 TO V + 1
580 P = P + A(I,V + 2) * (S(I) - T(I) * S(1) / N)
590    NEXT I
600 R = S(V + 2) - S(1) ^ 2 / N
610 Z = R - P
620 L = N - V - 1
630 I = P / V
```

```
640   PRINT
650 I = P / R
660   PRINT "COEFFICIENT OF ";
661   PRINT "DETERMINATION "
665   PRINT "                (R^2) =";I
670   PRINT "COEFFICIENT OF MULTIPLE"
675   PRINT "CORRELATION ="; SQR (I)
680   PRINT "STANDARD ERROR OF ESTIMATE = ";
681   PRINT  SQR ( ABS (Z / L))
690   PRINT
700   PRINT "INTERPOLATION: ";
701   PRINT "(ENTER 0 TO END PROGRAM)"
710 P = A(1,V + 2)
720   FOR J = 1 TO V
730   PRINT "VARIABLE";J;
740   INPUT X
750   IF X = 0 THEN 810
760 P = P + A(J + 1,V + 2) * X
770   NEXT J
780   PRINT "DEPENDENT VARIABLE =";P
790   PRINT
800   GOTO 710
810   PRINT : PRINT : PRINT "END OF PROGRAM"
820   END
```

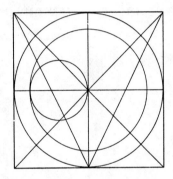

3
Data Analysis

Alphabetize

This program alphabetizes a list of words or phrases.

Numbers may be part of an alphanumeric phrase. However, they will not be put into numeric order unless they contain the same number of digits. Numbers with fewer digits must be justified to the right by prefixing zeros. Therefore, if the numbers you are sorting range into the hundreds, the number 13 would be entered as 013.

To save memory space, the array at statement 70 should be limited to the maximum number of terms you wish alphabetized. The dimension statement should be altered in the following manner:

$$70 \ DIM \ A\$(N)$$

where N = the number of items to be alphabetized.

Example

Alphabetize the following names:

Robert Wilson
Susan W. James
Kent Smith
Michael Mitchell
Ann T. McGowan
Alexander Lee II
Mary Mitchell
David Bowers
Steven Evans
Carol Jameson
Linda North

```
RUN
ALPHABETIZE

(TO END PROGRAM ENTER 0)
NUMBER OF ITEMS?11
ITEM1?WILSON ROBERT
ITEM2?JAMES SUSAN W.
ITEM3?SMITH KENT
ITEM4?MITCHELL MICHAEL
ITEM5?MCGOWAM ANN T.
ITEM6?LEE ALEXANDER II
ITEM7?MITCHELL MARY
ITEM8?BOWERS DAVID
ITEM9?EVANS STEVEN
ITEM10?JAMESON CAROL
ITEM11?NORTH LINDA
BOWERS DAVID
EVANS STEVEN
```

JAMES SUSAN W.
JAMESON CAROL
LEE ALEXANDER II
MCGOWAM ANN T.
MITCHELL MARY
MITCHELL MICHAEL
NORTH LINDA
SMITH KENT
WILSON ROBERT

ALPHABETIZE ANOTHER SET OF DATA?
(MUST CONTAIN THE SAME NUMBER OF ITEMS)
YES(1) OR NO(0)0
END OF PROGRAM

Program Listing

```
5    HOME
10   PRINT "ALPHABETIZE"
20   PRINT
30   PRINT "(TO END PROGRAM ENTER 0)"
40   PRINT "NUMBER OF ITEMS";
50   INPUT N
60   IF N = 0 THEN 330
70   DIM A$(N + 1)
80   FOR I = 1 TO N
90   PRINT "ITEM";I;
100   INPUT A$(I)
110   NEXT I
120  M = N
130  T = M / 2
131  M =   INT (T)
140   IF M = 0 THEN 280
150  K = N - M
160  J = 1
170  I = J
180  L = I + M
190   IF A$(I) <  = A$(L) THEN 250
200  T$ = A$(I)
210  A$(I) = A$(L)
220  A$(L) = T$
230  I = I - M
240   IF I >  = 1 THEN 180
250  J = J + 1
260   IF J > K THEN 130
270   GOTO 170
280   FOR I = 1 TO N
290   PRINT A$(I)
300   NEXT I
310   PRINT
320   PRINT "ALPHABETIZE ANOTHER SET OF DATA?"
322   PRINT "(MUST CONTAIN THE SAME NUMBER OF ITEMS)"
```

```
324    INPUT "YES(1) OR NO(0)";P
340    IF P = 1 THEN 400
350    PRINT "END OF PROGRAM"
355    END
400    HOME
405    PRINT "NUMBER OF ITEMS = ";N
410    PRINT : PRINT :
420    GOTO 80
```

Peak Finder

This program finds the location of peaks (and valleys) in a series of data points. By using the method of least squares, it provides an estimate of the true location and value of the peak.

Input data is a list of equally spaced data points; that is, the sampling interval must be constant. The program provides three modes of operation on the data. In mode 1, only the largest and smallest data values are selected. In mode 2, the program first computes a derivative at each point, then finds the zeros of the derivative corresponding to relative maxima and minima of the data. In mode 3, any X coordinate may be input from the keyboard.

In all modes, whatever point is chosen will be considered an approximate X coordinate of a local maximum or minimum.

Example

The following data was obtained by counting a radioactive tracer as it passed through the heart. Each point represents 25 milliseconds. Find the time at which the counting rate is the highest.

	1	2	3	4	5	6	7	8	9	10
Time (msec)	25	50	75	100	125	150	175	200	225	250
Count	84	85	83	105	108	123	107	96	102	92

	11	12	13	14	15	16	17	18	19	20
Time (msec)	275	300	325	350	375	400	425	450	475	500
Count	111	93	102	76	78	64	64	61	54	51

```
RUN
PEAK FINDER

THIS PROGRAM FINDS PEAKS AND VALLEYS
IN DATA CURVES AND PRINTS THEIR COORDINATES.

THE INPUT IS A LIST OF EQUALLY SPACED DATA POINTS.
THERE ARE THREE MODES OF OPERATION:
     1 = ABSOLUTE MAX AND MIN ONLY
     2 = ALL RELATIVE MAX AND MINS
     3 = APPROXIMATE COORDINATE
         ENTERED MANUALLY


ENTER NUMBER OF DATA POINTS: ?20
Y(1) = ?84
Y(2) = ?85
Y(3) = ?83
Y(4) = ?105
Y(5) = ?108
Y(6) = ?123
Y(7) = ?107
```

```
Y(8)  = ?96
Y(9)  = ?102
Y(10) = ?92
Y(11) = ?111
Y(12) = ?93
Y(13) = ?102
Y(14) = ?76
Y(15) = ?78
Y(16) = ?64
Y(17) = ?64
Y(18) = ?61
Y(19) = ?54
Y(20) = ?51


CHOOSE MODE:
     1 = ABS MAX AND MIN
     2 = RELATIVE MAX AND MINS
     3 = MANUAL ENTRY
WHICH ? ?2

     MAX = 116.442724 AT X = 5.77457627


     MIN = 95.3015585 AT X = 8.93636364


     MAX = 104.733889 AT X = 11.5444444

0 = ENTER NEW DATA
1 = REPEAT PEAKS
2 = STOP
CHOOSE ONE?2
END OF PROGRAM
```

Program Listing

```
100    HOME
170    PRINT "PEAK FINDER"
180    PRINT
190    PRINT "THIS PROGRAM FINDS PEAKS AND VALLEYS"
200    PRINT "IN DATA CURVES AND PRINTS THEIR COORDINATES."
210    PRINT
220    PRINT "THE INPUT IS A LIST OF EQUALLY SPACED DATA POINTS."
230    PRINT "THERE ARE THREE MODES OF OPERATION:"
240    PRINT "     1 = ABSOLUTE MAX AND MIN ONLY"
250    PRINT "     2 = ALL RELATIVE MAX AND MINS"
260    PRINT "     3 = APPROXIMATE COORDINATE"
265    PRINT "         ENTERED MANUALLY"
270    PRINT
280    PRINT
285    REM ==ENTER DATA==
290    DIM Y(100),Z(100)
300    PRINT "ENTER NUMBER OF DATA POINTS: ";
```

```
310    INPUT N
320    FOR K = 1 TO N
330    PRINT "Y(";K") = ";
340    INPUT Y(K)
350    NEXT K
360    PRINT
370    PRINT
379    REM ==CHOOSE MODE OF OPERATION==
380    PRINT "CHOOSE MODE:"
390    PRINT "       1 = ABS MAX AND MIN"
400    PRINT "       2 = RELATIVE MAX AND MINS"
410    PRINT "       3 = MANUAL ENTRY"
420    PRINT "WHICH ? ";
430    INPUT Q
440    IF Q < 1 THEN 420
445    IF Q = 1 THEN 540
450    IF Q = 2 THEN 750
455    IF Q = 3 THEN 1020
460    GOTO 380
469    REM ==REPEAT OR FINISH?==
470    PRINT "0 = ENTER NEW DATA"
480    PRINT "1 = REPEAT PEAKS"
490    PRINT "2 = STOP"
500    PRINT "CHOOSE ONE";
510    INPUT Q
515    IF Q = 0 THEN 300
520    IF Q = 1 THEN 380
525    PRINT "END OF PROGRAM"
530    END
531    REM ==END OF MAIN PROGRAM==
536    REM ==ABSOLUTE MIN AND MAX==
540 A = Y(3)
550 B = Y(3)
560 J = 3
570 K = 3
579    REM ==FIND MAX AND MIN==
580    FOR R = 3 TO N - 2
590    IF A > Y(R) THEN 630
600 A = Y(R)
610 J = R
620    GOTO 660
630    IF Y(R) > B THEN 660
640 B = Y(R)
650 K = R
660    NEXT R
669    REM ==DO FITTING AND PRINT RESULTS==
670    GOSUB 1110
690    GOSUB 1230
700 J = K
710    GOSUB 1110
730    GOSUB 1230
740    GOTO 470
745    REM ==RELATIVE MAX AN MIN==
749    REM ==FIND FIRST DERIVATIVE==
750    FOR K = 3 TO N - 2
```

```
760 Z(K) = Y(K + 1) - Y(K - 1) + 2 * (Y(K + 2) - Y(K - 2))
770 Z(K) = Z(K) / 10
780   NEXT K
790 Z(1) = Z(3)
800 Z(2) = Z(3)
810 Z(N) = Z(N - 2)
820 Z(N - 1) = Z(N)
825   REM
826   REM ==FIND ZEROS AND FIT EACH==
830   FOR K = 4 TO N - 2
840 T = Z(K) * Z(K - 1)
850   IF T =  > 0 THEN 1000
860 J = K
870   GOSUB 1110
880 U = X
890 J = K - 1
900   GOSUB 1110
910   IF  ABS (X) <  ABS (U) THEN 990
920 J = K
930   GOSUB 1110
990   GOSUB 1230
1000   NEXT K
1010   GOTO 470
1015   REM ==MANUAL ENTRY OF X==
1020   PRINT
1030   PRINT "ENTER X (1 TO STOP)";
1040   INPUT J
1050   IF J < 3 THEN 1100
1060   IF J > N - 2 THEN 1100
1070   GOSUB 1110
1080   GOSUB 1230
1090   GOTO 1020
1100   GOTO 470
1101   REM ==CALCULATE COEFFICIENTS OF PARABOLIC FIT==
1110 C = 17 * Y(J) + 12 * (Y(J-1) + Y(J+1)) - 3 * (Y(J-2) + Y(J+2))
1120 C = C / 35
1130 D = Y(J + 1) - Y(J - 1) + 2 * (Y(J + 2) - Y(J - 2))
1140 D = D / 10
1150 E = 2 * (Y(J - 2) - Y(J) + Y(J + 2)) - Y(J - 1) - Y(J + 1)
1160 E = E / 7
1170   IF E = 0 THEN 1210
1180 X =  - D / E
1190 P = C + (D + E * X / 2) * X
1191 B$ = "     MAX = "
1192   IF E < 0 THEN 1220
1193 B$ = "     MIN = "
1200   GOTO 1220
1210 B$ = "INDETERMINATE "
1211 X = 0
1212 P = 0
1220   RETURN
1221   REM ==PRINT RESULT==
1230   PRINT
1240   PRINT B$;P;" AT X = ";J + X
1250   PRINT
```

```
1260    RETURN
1270    END
```

References

Savitzky, A., and Golay, J. "Smoothing and Differentiation of Data by Simplified Least Squares
    Procedure." *Analytical Chemistry* 36 (1964), p. 1627.

Data Bounding, Smoothing, and Differentiation

This program offers two methods for reducing noise (invalid information) in a series of data points. This "noise filtering" is performed by either data bounding or by least squares smoothing. Four degrees of data smoothing are available (2-5) corresponding to either 5, 7, 9, or 11 points of data comparison in the program. It is also possible to use this program to calculate first and second derivatives. Proper weighting coefficients are given in the DATA statements for using from 5 to 11 data points.

Program Notes

In running this program you will be asked to select type of process:

$$-1 = \text{data bounding}$$
$$0 = \text{smoothing}$$
$$1 = \text{first derivative}$$
$$2 = \text{second derivative}$$

It will then be necessary to enter the degree of smoothing desired of your data. This is entered as a single digit from 2 to 5. Enter data in the following form:

$$X(1) = \text{DATA 1}$$
$$X(2) = \text{DATA 2}$$
$$\vdots \qquad \vdots$$
$$X(N) = \text{DATA } N$$

Example

For the following set of data points, compare the effects of data bounding, to various degrees of data smoothing. Notice how data bounding removes an aberrant point "cleanly," whereas smoothing simply spreads it out over several neighboring points.

84	123	111	64
85	107	93	64
83	96	102	61
105	102	76	54
108	92	78	51

```
RUN
DATA BOUNDING, SMOOTHING, AND
 DIFFERENTIATION

ENTER NUMBER OF DATA POINTS?20
X(1)=?84
X(2)=?85
X(3)=?83
X(4)=?105
X(5)=?108
```

```
X(6)=?123
X(7)=?107
X(8)=?96
X(9)=?102
X(10)=?92
X(11)=?111
X(12)=?93
X(13)=?102
X(14)=?76
X(15)=?78
X(16)=?64
X(17)=?64
X(18)=?61
X(19)=?54
X(20)=?51
SELECT TYPE OF PROCESS:
        -1 = DATA BOUNDING
         0 = SMOOTHING
         1 = FIRST DERIVATIVE
         2 = SECOND DERIVATIVE

ENTER CHOICE:?-1

DATA BOUNDING
HOW MANY STANDARD DEVIATIONS?1
RESULT: DATA BOUNDED
X(1)= 84          Y(1)= 84
X(2)= 85          Y(2)= 85
X(3)= 83          Y(3)= 92.1104336
X(4)= 105          Y(4)= 105
X(5)= 108          Y(5)= 108
X(6)= 123          Y(6)= 111.909464
X(7)= 107          Y(7)= 107
X(8)= 96          Y(8)= 96
X(9)= 102          Y(9)= 102
X(10)= 92          Y(10)= 101.591663
X(11)= 111          Y(11)= 100.464346
X(12)= 93          Y(12)= 102.643651
X(13)= 102          Y(13)= 91.9004951
X(14)= 76          Y(14)= 84.7177979
X(15)= 78          Y(15)= 78
X(16)= 64          Y(16)= 64
X(17)= 64          Y(17)= 64
X(18)= 61          Y(18)= 61
X(19)= 54          Y(19)= 54
X(20)= 51          Y(20)= 51
```

Program Listing

```
100    HOME
140    PRINT "DATA BOUNDING, SMOOTHING, AND"
145    PRINT " DIFFERENTIATION"
150    PRINT
235    REM ==DATA INPUT==
260    PRINT "ENTER NUMBER OF DATA POINTS";
```

```
270    INPUT N
280    FOR K = 1 TO N
290    PRINT "X(";K;")=";
300    INPUT X(K)
310    NEXT K
316    REM
317    REM
318    REM : PROCESSING MENU
319    REM
320    PRINT "SELECT TYPE OF PROCESS:"
330    PRINT "        -1 = DATA BOUNDING"
340    PRINT "         0 = SMOOTHING"
350    PRINT "         1 = FIRST DERIVATIVE"
360    PRINT "         2 = SECOND DERIVATIVE"
370    PRINT
380    PRINT "ENTER CHOICE:";
390    INPUT Q
400    IF Q <  - 1 THEN 440
410    IF Q =  - 1 THEN 520
420    IF Q > 2 THEN 440
430    GOTO 720
440    PRINT "ANSWER MUST BE IN RANGE -1 TO 2"
450    GOTO 370
456    REM ==PRINT RESULTS==
460    PRINT "RESULT: ";A$
470    FOR K = 1 TO N
480    PRINT "X(";K;")= ";X(K);"            ";
490    PRINT "Y(";K;")= ";Y(K)
500    NEXT K
510    GOTO 1800
511    REM ==DATA BOUNDING ROUTINE==
520    PRINT
530    PRINT "DATA BOUNDING"
540  A$ = "DATA BOUNDED"
550    PRINT "HOW MANY STANDARD DEVIATIONS";
560    INPUT S
570  Y(1) = X(1)
580    FOR K = 2 TO N - 1
590  A = (X(K - 1) + X(K + 1)) / 2
600  D = X(K) - A
610  M = S *  SQR (X(K))
620    IF  ABS (D) <  ABS (M) THEN 680
630    IF D > 0 THEN 660
640  Y(K) = X(K) + M
650    GOTO 690
660  Y(K) = X(K) - M
670    GOTO 690
680  Y(K) = X(K)
690    NEXT K
700  Y(N) = X(N)
710    GOTO 460
711    REM ==LEAST SQUARES SMOOTHING ROUTINE==
720    PRINT
730    PRINT "DEGREE OF SMOOTHING DESIRED (2-4)
740    INPUT S
```

```
750    IF S < 2 THEN 780
760    IF S > 4 THEN 780
770    GOTO 800
780    PRINT "ANSWER MUST BE IN THE RANGE (2-4)"
790    GOTO 720
797    REM ==SET POINTER AND READ CONVOLUTION ARRAY=
800    RESTORE
810 T = 36 * Q + (S - 2) * (S + 3)
820    FOR K = 1 TO T + 1
830    READ B
840    NEXT K
850 M = 2 * S + 1
860    FOR K = 1 TO M
870    READ C(K)
880    NEXT K
890 M1 = M
897    REM ==SMOOTH OR DIFFERENTIATE EACH DATA POINT==
900    FOR K = S + 1 TO N - S
910 D = 0
920 L = M1
927    REM ==COMPUTE CONVOLUTION==
930    FOR J = 1 TO M
940 D = D + X(L) * C(J)
950 L = L - 1
960    NEXT J
970 Y(K) = D / 8
980 M1 = M1 + 1
990    NEXT K
996    REM ==CONVOLUTION ARRAY COEFFICIENTS==
1111   REM ==COMPUTE END POINTS, SECOND DERIVATIVE FIRST==
1120   FOR K = 1 TO S
1130 Y(K) = Y(S + 1)
1140   NEXT K
1150   FOR K = N - S + 1 TO N
1160 Y(K) = Y(N - S)
1170   NEXT K
1180 A$ = "2ND DERIVATIVE"
1190   IF Q = 2 THEN 1720
1197   REM ==FIRST DERIVATIVE==
1200 T = 35 - 2 * S
1201   REM ==READ NEW SET OF COEFFICIENTS==
1210   FOR K = 1 TO T
1220   READ B
1230   NEXT K
1240   FOR K = 1 TO M
1250   READ C(K)
1260   NEXT K
1270 D = 0
1280 L = N
1290 R = M
1300 E = 0
1301   REM :COMPUTE CONVOLUTION
1310   FOR J = 1 TO M
1320 E = E + X(L) * C(J)
1330 D = D + X(R) * C(J)
```

```
1340 R = R = 1
1350 L = L - 1
1360   NEXT J
1370 D = D / B
1380 E = E / B
1390 L = N - S + 1
1400   FOR K = 1 TO S
1410 Y(K) = Y(K) + D * (K - S - 1)
1420 Y(L) = Y(L) + E * K
1430 L = L + 1
1440   NEXT K
1450 A$ = "1ST DERIVATIVE"
1460   IF Q = 1 THEN 1720
1461   REM ==SMOOTHED VALUES==
1469   REM ==READ NEW COEFFICIENTS==
1470   FOR K = 1 TO T
1480   READ B
1490   NEXT K
1500   FOR K = 1 TO M
1510   READ C(K)
1520   NEXT K
1530 D = 0
1540 E = 0
1550 R = M
1560 L = N
1569   REM ==COMPUTE CONVOLUTION==
1570   FOR J = 1 TO M
1580 D = D + X(R) * C(J)
1590 E = E + X(L) * C(J)
1600 R = R - 1
1610 L = L - 1
1620   NEXT J
1630 D = D / B
1640 E = E / B
1650 L = N - S + 1
1660   FOR K = 1 TO S
1670 Y(K) = Y(K) + D * (K - S - 1) * (K - S - 1) / 2
1680 Y(L) = Y(L) + E * K * K / 2
1690 L = L + 1
1700   NEXT K
1710 A$ = "SMOOTHED DATA"
1720   GOTO 460
1800   END
2000   DATA 35,-3,12,17,12,-3
2010   DATA   21,-2,3,6,7,6,3,-2
2030   DATA     231,-21,14,39,54,59,54,39,14,-21
2040   DATA     429,-36,9,44,69,84,89,84,69,44,9,-36
2050   DATA     10,2,1,0,-1,-2
2060   DATA     28,3,2,1,0,-1,-2,-3
2070   DATA     60,4,3,2,1,0,-1,-2,-3,-4
2080   DATA     110,5,4,3,2,1,0,-1,-2,-3,-4,-5
2090   DATA   7,2,-1,-2,-1,2
2100   DATA   42,5,0,-3,-4,-3,0,5
2110   DATA     462,28,7,-8,-17,-8,7,28
2120   DATA     429,15,6,-1,-9,-10,-9,-6,-1,6,15
```

References

Diffey, B. L., and Corfield, J. R. "Data-Bounding Technique in Discrete Deconvolution." *Medical and Biological Engineering,* vol. 14 (1976).

Savitzky, A., and Golay, J. "Smoothing and Differentiation of Data by Simplified Least Squares Procedure." *Analytical Chemistry* 36 (1964), p. 1627.

Convolution/Deconvolution

This program provides a simple routine to compute convolutions and deconvolutions. The input to the program consists of two "curves," or lists of data values. The data must be obtained by sampling at equal intervals. Convolution is achieved by entering an input function $B(J)$ and a response function $A(J)$, where $J = I$ to N indexes the data by time or space. The equation solved for convolution is:

$$C \ (N) = \sum_{J=1}^{N} B(J) \times A(N-J+1)$$

for $I = 1,N$. Note that the upper limit of summation is I. This preserves causality; that is, there can be no output before there is input.

Program Notes

Deconvolution is achieved by a simple algebraic solution of the convolution equation. Suppose we want to solve for $A(I)$. The first convolution equation, for $I = 1$, is simply:

$$C(1) = B(1) \times A(1)$$

This may be inverted to give $A(1)$. Each subsequent equation may be solved in order, since each $A(N)$ depends only on $A(J)$ for $J < N$. If $C(N)$ and $A(N)$ are known, $B(N)$ may be found in the same way.

Example

An experiment was tried to see if the usage of natural gas for heating could be predicted from the temperature recorded as heating degree-days. The monthly degree-days recorded for Baltimore for 1978 were entered as curve B. The amount of gas billed in those months was entered as curve C. Since the billing month did not coincide with the calendar month, there was some lag between the temperature and the gas usage. The deconvoluted curve, or "response function," is computed as curve A.

To test the hypothesis and the program, the heating degree-days for 1979 were entered as curve B. The convolution curve C then gives the predicted amounts used for 1979. For comparison, the actual usage is given along with a prediction based on linear regression.

```
RUN

CONVOLUTION/DECONVOLUTION


ENTER NUMBER OF POINTS/CURVE:?12

0=STOP
1=ENTER CURVE A
2=ENTER CURVE B
3=ENTER CURVE C
4=CONVOLUTION: C=A*B
```

```
5=DECONVOLUTION: A=C/B
6=DECONVOLUTION: B=C/A
     WHICH??2

ENTER VALUES OF CURVE ´B´:
  B(1)  = ?1101
  B(2)  = ?1048
  B(3)  = ?715
  B(4)  = ?318
  B(5)  = ?141
  B(6)  = ?9
  B(7)  = ?0
  B(8)  = ?0
  B(9)  = ?33
  B(10) = ?280
  B(11) = ?483
  B(12) = ?763


0=STOP
1=ENTER CURVE A
2=ENTER CURVE B
3=ENTER CURVE C
4=CONVOLUTION: C=A*B
5=DECONVOLUTION: A=C/B
6=DECONVOLUTION: B=C/A
     WHICH??3

ENTER VALUES OF CURVE ´C´:
  C(1)  = ?173
  C(2)  = ?215
  C(3)  = ?196
  C(4)  = ?93
  C(5)  = ?73
  C(6)  = ?41
  C(7)  = ?32
  C(8)  = ?27
  C(9)  = ?29
  C(10) = ?29
  C(11) = ?71
  C(12) = ?125


0=STOP
1=ENTER CURVE A
2=ENTER CURVE B
3=ENTER CURVE C
4=CONVOLUTION: C=A*B
5=DECONVOLUTION: A=C/B
6=DECONVOLUTION: B=C/A
     WHICH??5

A(1) = .157129882
A(2) = .0457110661
A(3) = .0324676995
```

```
A(4)  =  -.0215048718
A(5)  =  .0323626413
A(6)  =  3.88353996E-03
A(7)  =  6.03084974E-03
A(8)  =  9.40199462E-03
A(9)  =  3.67376173E-03
A(10) =  -.0270971968
A(11) =  2.84443333E-03
A(12) =  -.0104491603


0=STOP
1=ENTER CURVE A
2=ENTER CURVE B
3=ENTER CURVE C
4=CONVOLUTION:  C=A*B
5=DECONVOLUTION:  A=C/B
6=DECONVOLUTION:  B=C/A
     WHICH??2


ENTER VALUES OF CURVE 'B':
  B(1)  = ?984
  B(2)  = ?1100
  B(3)  = ?520
  B(4)  = ?354
  B(5)  = ?75
  B(6)  = ?6
  B(7)  = ?2
  B(8)  = ?3
  B(9)  = ?22
  B(10) = ?311
  B(11) = ?425
  B(12) = ?757



0=STOP
1=ENTER CURVE A
2=ENTER CURVE B
3=ENTER CURVE C
4=CONVOLUTION:  C=A*B
5=DECONVOLUTION:  A=C/B
6=DECONVOLUTION:  B=C/A
     WHICH??4


C(1)  = 154.615804
C(2)  = 217.822559
C(3)  = 163.937928
C(4)  = 93.9474083
C(5)  = 53.0391424
C(6)  = 44.1024504
C(7)  = 22.4457026
C(8)  = 28.5060658
C(9)  = 24.4250855
C(10) = 34.8143257
C(11) = 60.4171245
C(12) = 128.901933
```

```
0=STOP
1=ENTER CURVE A
2=ENTER CURVE B
3=ENTER CURVE C
4=CONVOLUTION: C=A*B
5=DECONVOLUTION: A=C/B
6=DECONVOLUTION: B=C/A
     WHICH??0

END OF PROGRAM
```

Program Listing

```
5    HOME
210   DIM A(100),B(100),C(100),T(3)
220   PRINT
230   PRINT "CONVOLUTION/DECONVOLUTION"
240   PRINT
300   PRINT
310   PRINT "ENTER NUMBER OF POINTS/CURVE:";
320   INPUT N
330 T(1) = 0
340 T(2) = 0
350 T(3) = 0
360   PRINT
370   PRINT "0=STOP"
380   PRINT "1=ENTER CURVE A"
390   PRINT "2=ENTER CURVE B"
400   PRINT "3=ENTER CURVE C"
410   PRINT "4=CONVOLUTION: C=A*B"
420   PRINT "5=DECONVOLUTION: A=C/B"
430   PRINT "6=DECONVOLUTION: B=C/A"
440   PRINT "     WHICH?";
450   INPUT Q
460   IF Q > 0 THEN 501
470   PRINT
480   PRINT "END OF PROGRAM"
490   END
501   IF Q = 1 THEN 520
502   IF Q = 2 THEN 610
503   IF Q = 3 THEN 700
504   IF Q = 4 THEN 790
505   IF Q = 5 THEN 930
506   IF Q = 6 THEN 1190
510   GOTO 360
511   REM ==END OF MAIN PROGRAM==
518   REM ==INPUT CURVE A==
520   PRINT
530   PRINT "ENTER VALUES OF CURVE 'A':"
540   FOR K = 1 TO N
550   PRINT "  A(";K;") = ";
560   INPUT A(K)
570   NEXT K
580   PRINT
```

```
590  T(1) = 1
600   GOTO 360
601   REM ==INPUT CURVE B==
610   PRINT
620   PRINT "ENTER VALUES OF CURVE 'B':"
630   FOR K = 1 TO N
640   PRINT "  B(";K;") = ";
650   INPUT B(K)
660   NEXT K
670   PRINT
680  T(2) = 1
690   GOTO 360
691   REM ==INPUT CURVE C==
700   PRINT
710   PRINT "ENTER VALUES OF CURVE 'C':"
720   FOR K = 1 TO N
730   PRINT "   C(";K;") = ";
740   INPUT C(K)
750   NEXT K
760   PRINT
770  T(3) = 1
780   GOTO 360
781   REM ==CONVOLUTION==
790   IF T(1) = 0 THEN 1450
800   IF T(2) = 0 THEN 145
801   REM ==COMPUTE CONVOLUTION==
810   FOR K = 1 TO N
820  C(K) = 0
830   FOR J = 1 TO K
840  C(K) = C(K) + A(J) * B(K - J + 1)
850   NEXT J
860   NEXT K
870   PRINT
879   REM ==PRINT RESULT==
880   FOR K = 1 TO N
890   PRINT "C(";K;") = ";C(K)
900   NEXT K
910  T(3) = 1
920   GOTO 360
921   REM ==DECONVOLUTION (MODE 5)==
930   IF T(2) = 0 THEN 1450
940   IF T(3) = 0 THEN 1450
950  M = B(1)
960  H = 1
969   REM ==FIND MAX OF B==
970   FOR K = 2 TO N
980   IF B(K) <  = M THEN 1010
990  M = B(K)
1000 H = K
1010   NEXT K
1020   FOR K = 1 TO H - 1
1030 A(K) = 0
1040   NEXT K
1050 A(H) = C(H) / B(H)
1055   IF H = N THEN 1140
```

```
1059   REM ==COMPUTE DECONVOLUTION==
1060   FOR K = H + 1 TO N
1070 A(K) = C(K)
1080   FOR J = H TO K - 1
1090 A(K) = A(K) - A(J) * B(K - J + H)
1100   NEXT J
1110 A(K) = A(K) / B(H)
1120   NEXT K
1130   PRINT
1139   REM ==PRINT RESULT==
1140   FOR K = 1 TO N
1150   PRINT "A(";K;") = ";A(K)
1160   NEXT K
1170 T(1) = 1
1180   GOTO 360
1190   END
```

References

Diffey, B. L.; Hall, F. M.; and Corfield, J. R. "The Tc-DTPA Dynamic Renal Scan with Deconvolution Analysis." *Journal of Nuclear Medicine* 17 (1976), pp. 352-55.

Valentinuzzi, M. E., and Montaldo Volachec, E. M. "Discrete Deconvolution." *Medical and Biological Engineering,* vol. 13 (1975), pp. 123-25.

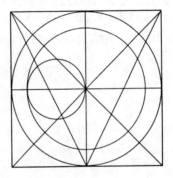

4
Roots of Polynomials

Roots of Polynomials:
Half-Interval Search

Real Roots of Polynomials:
Newton

Roots of Polynomials:
Bairstow's Method

Roots of the General Polynomial

Roots of Polynomials: Half-Interval Search

This program calculates roots of polynomials within a given interval. The program first conducts a random search within the given interval for two points with opposite signs. If a change of sign is found, then the root is calculated by the half-interval search method. If there is no change of sign found, another interval will be asked for.

Errors may result in this program for two reasons. First, a root may be calculated when it should not be. This may happen if the lowest point is so close to zero that a root is found due to round-off error. Second, two roots may be so close together that the program never finds the opposite signs between them. The result in this case is that neither root is calculated.

It is necessary to enter the equation before you run the program. The equation will be defined as a function of X at statement 30. For example, if you want to find roots of the function $f(X) = 4X^4 - 2.5X^2 - X + 0.5$, you will enter:

```
30 DEFFNR(X)=4*X^4-2.5*X^2-X+.5
```

Example

Find a root of the function $f(X) = 4X^4 - 2.5X^2 - X + 0.5$.

```
DEFFNR(X)=4*X^4-2.5*X^2-X+0.5

RUN
ROOTS OF POLYNOMIALS
BY HALF-INTERVAL SEARCH


(TO END SEARCH ENTER 0,0)

INTERVAL (LOWER,UPPER)?-1,0


       - SEARCHING FOR ROOT -


NO CHANGE OF SIGN FOUND
INTERVAL (LOWER,UPPER)?0,1


       - SEARCHING FOR ROOT -


ROOT =.303576073

INTERVAL (LOWER,UPPER)?0,0
```

Program Listing

```
5   HOME
20    PRINT "ROOTS OF POLYNOMIALS"
```

```
25    PRINT "BY HALF-INTERVAL SEARCH"
26    PRINT : PRINT
29    REM ==ENTER FUNCTION HERE==
30    DEF   FN R(X) = 4 * X ^ 4 - 2.5 * X ^ 2 - X + .5
50    PRINT "(TO END SEARCH ENTER 0,0)"
55    PRINT
59    REM ==ESTABLISH INTERVAL OF RANDOM SEARCH==
60    PRINT "INTERVAL (LOWER,UPPER)";
70    INPUT A,B
79    REM ==TEST FOR USABLE LIMITS ENTERED==
80    IF A <  > B THEN 120
90    IF A = 0 THEN 430
100    PRINT "--INTERVAL LIMITS CANNOT";
105    PRINT "BE EQUAL--"
110    GOTO 60
120    IF A < B THEN 150
130    PRINT "--LOWER LIMIT MUST BE ";
135    PRINT "ENTERED FIRST--"
140    GOTO 60
150 A1 =  SGN ( FN R(A))
160 B1 =  SGN ( FN R(B))
169    REM ==TEST FOR ROOT AT EITHER LIMIT==
170    IF A1 * B1 = 0 THEN 360
179    REM ==TEST FOR OPPOSITE SIGNS AT INTERVAL LIMITS==
180    IF A1 * B1 < 0 THEN 280
184    PRINT : PRINT
185    PRINT "          - SEARCHING FOR ROOT -"
186    PRINT : PRINT
189    REM ==LOOP TO SEARCH 1000 NUMBERS FOR OPPOSITE SIGNS IN FUNCTION==
190    FOR I = 1 TO 1000
200 X = A +  RND (2) * (B - A)
210 X1 =  SGN ( FN R(X))
219    REM ==TEST FOR ROOT AT RANDOM NUMBER; IF YES,END SEARCH,PRINT==
220    IF X1 = 0 THEN 400
229    REM ==TEST FOR OPPOSITE SIGNS AT RANDOM NUMBER AND LOWER LIMIT==
230    IF A1 * X1 < 0 THEN 270
239    REM ==TRY ANOTHER RANDOM NUMBER==
240    NEXT I
250    PRINT "NO CHANGE OF SIGN FOUND"
260    GOTO 60
269    REM ==CHANGE OF SIGN FOUND, CALCULATE ROOT==
270 B = X
278    REM ==STORE POSITIVE POINT IN D(3), NEGATIVE POINT IN D(1)==
279    REM  * D(1) AND D(3) BECOME INTERVAL LIMITS *
280 D(2 + A1) = A
290 D(2 - A1) = B
299    REM ==CALCULATE MIDPOINT BETWEEN THE TWO LIMITS==
300 Y = (D(1) + D(3)) / 2
310 Y1 =  SGN ( FN R(Y))
319    REM ==TEST FOR ROOT AT MIDPOINT==
320    IF Y1 = 0 THEN 400
329    REM ==GET A NEW LIMIT TO CLOSE IN ON ROT==
330 D(2 + Y1) = Y
339    REM ==TEST FOR A VALUE CLOSE ENOUGH TO A Q TO ASSUME A ROOT==
340    IF  ABS (D(1) - D(3)) /  ABS (D(1) +  ABS (D(3))) < 5E - 6 THEN 400
349    REM ==RETEST W/ NEW LIMITS==
```

```
350    GOTO 300
359    REM ==ROOT AT AN INTERVAL LIMIT; FIND WHICH LIMIT, PRINT==
360    IF A1 = 0 THEN 390
370 Y = B
380    GOTO 400
390 Y = A
400    PRINT "ROOT =";Y
410    PRINT
419    REM ==RESTART PROGRAM==
420    GOTO 60
430    END
```

Real Roots of Polynomials: Newton

This program calculates real roots of a polynomial with real coefficients. You must give an estimate of each root.

The calculations are performed using Newton's method for approximating roots of equations. The value of the error and derivative are included for each root calculated.

The equation you enter is presently limited to a degree of 10. You may enter a larger degree of equation by altering statements 30 and 40 of the program according to the following scheme:

```
30 DIM A(N+1), B(N+1)
40 FOR I = 1 TO N+1
```

where N = degree of equation.

Example

Find the roots of $4 X^4 - 2.5 X^2 - X + 0.5$.

```
RUN
REAL ROOTS OF POLYNOMIALS

NEWTON

DEGREE OF EQUATION?4

COEFFICIENT A(0)?.5
COEFFICIENT A(1)?-1
COEFFICIENT A(2)?-2.5
COEFFICIENT A(3)?0
COEFFICIENT A(4)?4

GUESS?-.8
            ROOT            ERROR            DERIVATIVE
       .30357634      -2.91038305E-11      -2.070247

NEW VALUE (1=YES, 0=NO)?0
NEW FUNCTION (1=YES, 0=NO)?0
```

Program Listing

```
5   HOME
10    PRINT "REAL ROOTS OF POLYNOMIALS"
12    PRINT
15    PRINT "NEWTON"
20    PRINT
30    DIM A(11),B(11)
40    FOR I = 1 TO 11
50  A(I) = 0
```

```
60 B(I) = 0
70   NEXT I
80   PRINT "DEGREE OF EQUATION";
90   INPUT N
95   PRINT
100  FOR I = 1 TO N + 1
110  PRINT "COEFFICIENT A(";I - 1;")";
120  INPUT A(I)
130  NEXT I
140  FOR I = 1 TO 10
150 B(I) = A(I + 1) * I
160  NEXT I
170  PRINT
180  PRINT "GUESS";
190  INPUT X
200 Q = 0
210 S = 1
220 F1 = 0
230 F0 = 0
240 Q = Q + 1
250  FOR I = 1 TO N + 1
260 F0 = F0 + A(I) * S
270 F1 = F1 + B(I) * S
280 S = S * X
290  NEXT I
300  IF F1 = 0 THEN 360
310 S = X - F0 / F1
320  IF X = S THEN 380
330 X = S
340  IF Q > 100 THEN 490
350  GOTO 210
360  PRINT "DERIVATIVE = 0 AT X =";X
370  GOTO 180
380  PRINT
390  PRINT " ROOT"; TAB( 13);"ERROR";
395  PRINT  TAB( 29);"DERIVATIVE"
400  PRINT X; TAB( 13);F0; TAB( 29);F1
410  PRINT
420  PRINT "NEW VALUE (1=YES, 0=NO)";
430  INPUT A
440  IF A = 1 THEN 170
450  PRINT "NEW FUNCTION (1=YES, 0=NO)";
460  INPUT A
470  IF A = 1 THEN 40
480  GOTO 550
490  PRINT "100 ITERATIONS COMPLETED:"
500  PRINT "X =";X;"F(X) =";F0
510  PRINT "  CONTINUE (1=YES,0=NO)";
520  INPUT A
530  IF A = 1 THEN 200
540  GOTO 420
550  END
```

Roots of Polynomials: Bairstow's Method

This program will find the real and complex roots of an Nth degree polynomial with real coefficients by Bairstow's algorithm:

$$\text{for:} \quad (n < 20)$$
$$P(X) = A_0 + A_1 + \ldots + A_n X^n$$

Example

```
RUN
ROOTS OF POLYNOMIALS (BAIRSTOWS METHOD)

DEGREE OF POLYNOMIAL?4

ENTER COEFFICIENTS:
A(0)?12
A(1)?-19
A(2)?12
A(3)?-6
A(4)?1

ROOTS:

.499999993+ I *1.6583124
.499999993- I *1.6583124


4
1.00000002

WOULD YOU LIKE TO RE-RUN THIS PROGRAM WITH NEW DATA (Y/N)?Y

DEGREE OF POLYNOMIAL?4

ENTER COEFFICIENTS:
A(0)?3.3
A(1)?.5
A(2)?2.3
A(3)?-1.1
A(4)?1

ROOTS:

-.45+ I *.947364767
-.45- I *.947364767

1+ I *1.41421356
1- I *1.41421356

WOULD YOU LIKE TO RE-RUN THIS PROGRAM WITH NEW DATA (Y/N)?N
```

Program Listing

```
5    HOME
10   PRINT "ROOTS OF POLYNOMIALS (BAIRSTOWS METHOD)"
30   DIM A(22),B(22),E(22)
40 E1 = 1E - 4
50 E4 = 1E - 20
60 K1 = 100
70   PRINT
80   PRINT "DEGREE OF POLYNOMIAL";
90   INPUT N
100   PRINT
120   PRINT "ENTER COEFFICIENTS:"
130   FOR I = 0 TO N
140   PRINT "A(";I;")";
150   REM ===COEFFICIENTS ARE STORED IN ARRAY A() IN REVERSE ORDER==
160   INPUT A(N - I + 1)
170   NEXT I
180   REM ==TEST FOR INVALID DATA==
190   IF A(1) <  > 0 THEN 270
200   REM ==PROMPT FOR AN INVALID ENTRY==
210   PRINT "A(N) MUST BE NON-ZERO.  RE-ENTER."
220   PRINT "A(";N;")";
230   INPUT A(1)
240   REM ==TEST NEW DATA==
250   GOTO 190
260   REM
270   PRINT
280   PRINT "ROOTS:"
290   REM ==BRANCH FOR SPECIAL TREATMENT OF 1ST AND 2ND DEGREE EQUATIONS==
300   IF N <  = 2 THEN 1080
310 A(N + 2) = 0
320 N1 = 2 *  INT ((N + 1) / 2)
330   FOR M1 = 1 TO N1 / 2
340 P = 1
350 Q = 1
360   FOR K = 1 TO K1
370   FOR L = 1 TO K1
380   REM ==STORE ALL COEFFICIENTS IN ARRAY B==
390   FOR I = 1 TO N1 + 1
400 B(I) = A(I)
410   NEXT I
420   FOR J = N1 - 2 TO N1 - 4 STEP  - 2
430   FOR I = 1 TO J + 1
440 B(I + 1) = B(I + 1) - P * B(I)
450 B(I + 2) = B(I + 2) - Q * B(I)
460   NEXT I
470   NEXT J
480 R0 = B(N1 + 1)
490 R1 = B(N1)
500 S0 = B(N1 - 1)
510 S1 = B(N1 - 2)
520 V0 =   - Q * S1
530 V1 = S0 - S1 * P
540 D0 = V1 * S0 - V0 * S1
```

```
550    IF  ABS (D0) >  = E4 THEN 590
560 P = P + 5
570 Q = Q + 5
580    NEXT L
590 D1 = S0 * R1 - S1 * R0
600 D2 = R0 * V1 - V0 * R1
610 P1 = D1 / D0
620 Q1 = D2 / D0
630 P = P + P1
640 Q = Q + Q1
650    IF  ABS (R0) >  = E1 THEN 690
660    IF  ABS (R1) >  = E1 THEN 690
670 E(M1) = 1
680    GOTO 810
690    IF  ABS (P1) >  = E1 THEN 730
700    IF  ABS (Q1) >  = E1 THEN 730
710 E(M1) = 2
720    GOTO 810
730    IF P = 0 THEN 750
740    IF  ABS (P1 / P) >  = E1 THEN 790
750    IF Q = 0 THEN 790
760    IF  ABS (Q1 / Q) >  = E1 THEN 790
770 E(M1) = 3
780    GOTO 810
790    NEXT K
800 E(M1) = 4
810 S =  - P / 2
820 T = S * S - Q
830    IF T < 0 THEN 890
840 T =  SQR (T)
850    PRINT
860    PRINT S + T
870    PRINT S - T
880    GOTO 930
890 T =  SQR ( - T)
900    PRINT
910    PRINT S;"+ I *";T
920    PRINT S;"- I *";T
930    IF E(M1) = 4 THEN 1180
940    FOR J = 1 TO N1 - 1
950 A(J + 1) = A(J + 1) - P * A(J)
960 A(J + 2) = A(J + 2) - Q * A(J)
970    NEXT J
980 N1 = N1 - 2
990    IF N1 > 1 THEN 1010
1000    GOTO 1180
1010    IF N1 >  = 3 THEN 1070
1020 M1 = M1 + 1
1030 E(M1) = 1
1040 P = A(2) / A(1)
1050 Q = A(3) / A(1)
1060    GOTO 810
1070    NEXT M1
1080    IF N = 2 THEN 1110
1090    PRINT  - A(2) / A(1)
```

```
1100   GOTO 1180
1110 A(3) = A(2) * A(2) - 4 * A(1) * A(3)
1120 S =   - A(2) / 2 / A(1)
1130 T =   SQR ( ABS (A(3))) / 2 / A(1)
1140 M1 = 4
1150 E(4) = 4
1160   IF   SGN (A(3)) < 0 THEN 900
1170   GOTO 850
1180   PRINT
1190   PRINT "WOULD YOU LIKE TO RE-RUN THIS PROGRAM WITH NEW DATA (Y/N)";
1200   INPUT X$
1210   IF X$ = "Y" THEN 70
1220   IF X$ <  > "N" THEN 1190
1230   END
```

Roots of the General Polynomial

This program will compute the roots of a general polynomial using the method of real coefficients.

The coefficients of the polynomial are input from the zeroth degree to the N th degree, with each coefficient C_k keyed in as A_k, B_k. Any set of X, Y can be used to initialize, but it is recommended you seek out the smaller roots first, using 1,1 for example.

Program Notes

If all the conjugate pairs are real numbers, then the complex roots will occur in conjugate pairs (i.e., $3 + 3I$; $2 - 3I$). When one such root is obtained, then its conjugate can be used in the next initialization.

If, however, there are complex coefficients, this conjugate principle no longer holds. If there are repeated (multiple) roots, the process slows down noticeably. It will tend to "hunt" on either side of the true root. In this instance you can either fire up with new values, or simply have the root displayed at some stage.

Example

Find the root of:

$$7X^5 - 42X^4 + 716X^3 - 1017X^2 + 19X - 2073$$

```
ROOTS OF THE GENERAL POLYNOMIAL


     KEY IN DEGREE OF POLYNOMIAL 4
     KEY IN COMPONENTS A,B OF C() BY INDICATED POWER

             C(0) = ?-1773,4208
             C(1) = ?0,472
             C(2) = ?163,0
             C(3) = ?-30,2
             C(4) = ?2,3


     KEY IN GUESS OF X AND Y 1,1
FOR N = 4   ROOT IS (3.36041803,-5.18123113)
     KEY IN GUESS OF X AND Y 1,1
FOR N = 3   ROOT IS (-3.21778099,1.24583832)
     KEY IN GUESS OF X AND Y 1,1
FOR N = 2   ROOT IS (4.5961909,5.2250881)
FOR N = 1   ROOT IS (-.584981794,-8.52046453)


***  ALL ROOTS FOUND   ***
```

Program Listing

```
10    HOME
20    DIM A(10,10),B(10,10),P(10),C1(10)
30    PRINT "ROOTS OF THE GENERAL POLYNOMIAL"
40    PRINT
70    PRINT
80    PRINT : INPUT "              KEY IN ORDER ";N
90    PRINT "          KEY IN ELEMENTS AS INDICATED": PRINT
100   FOR I = 1 TO N
110   FOR J = 1 TO N
120   PRINT "     A(";I;",";J;") =";
130   INPUT A(I,J)
140 B(I,J) = A(I,J)
150   NEXT J
160   PRINT
170   NEXT I
180 M = N - 1
190   FOR K = 1 TO M
200 T1 = 0
210   FOR I = 1 TO N
220 T1 = T1 + B(I,I)
230   NEXT I
240 A1 = K
250 P(K) = T1 / A1
260   FOR I = 1 TO N
270 B(I,I) = B(I,I) - P(K)
280   NEXT I
290   FOR J = 1 TO N
300   FOR I = 1 TO N
310 C1(I) = B(I,J)
320   NEXT I
330   FOR I = 1 TO N
340 B(I,J) = 0
350   FOR L = 1 TO N
360 B(I,J) = B(I,J) + A(I,L) * C1(L)
370   NEXT L
380   NEXT I
390   NEXT J
400   NEXT K
410 P(N) = B(N,N)
420   HOME
430   PRINT "        <<<<< COEFFICIENTS ARE >>>>>"
440   PRINT
450   FOR K = 1 TO N
460   PRINT  TAB( 25)"P(";K;") = ";P(K)
470   NEXT K
480   END
```

References

"A Convergent Algorithm for Solving Polynomial Equations." *Journal AOM* vol. 14, #2, April 1967, pp. 311-15.

5
Linear Equations

Gaussian Elimination Method

Compact Crout Method

General Linear Equation Solver

Gaussian Elimination Method

This program solves a system of linear equations. The number of unknown coefficients in each equation must equal the number of equations being solved. You must enter the coefficients of each equation.

The dimension statement at line 30 limits the number of equations which may be solved. You may change this limit according to the following scheme:

$$30 \ \text{DIM} \ A(R, R+1)$$

where R = the maximum number of equations.

Example

Solve the following system of equations:

$$X_1 + 2X_2 + 3X_3 = 4$$
$$3X_1 + 6X_2 = 1$$
$$-3X_1 + 4X_2 - 2X_3 = 0$$

```
]30 DIM A(3,4)

]RUN
GAUSSIAN ELIMINATION METHOD

NUMBER OF EQUATIONS?3
COEFFICIENT MATRIX:

EQUATION1
     COEFFICIENT1?1
     COEFFICIENT2?2
     COEFFICIENT3?3
CONSTANT?4

EQUATION2
     COEFFICIENT1?3
     COEFFICIENT2?6
     COEFFICIENT3?0
CONSTANT?1

EQUATION3
     COEFFICIENT1?-3
     COEFFICIENT2?4
     COEFFICIENT3?-2
CONSTANT?0

X1=-.356
X2=.344
X3=1.222
```

Program Listing

```
5    HOME
10   PRINT "GAUSSIAN ELIMINATION METHOD"
20   PRINT
30   DIM A(15,15)
40   PRINT "NUMBER OF EQUATIONS";
50   INPUT R
60   PRINT "COEFFICIENT MATRIX:"
70   FOR J = 1 TO R
80   PRINT
85   PRINT "EQUATION";J
90   FOR I = 1 TO R + 1
100   IF I = R + 1 THEN 130
110   PRINT "     COEFFICIENT";I;
120   GOTO 140
130   PRINT "CONSTANT";
140   INPUT A(J,I)
150   NEXT I
160   NEXT J
170   FOR J = 1 TO R
180   FOR I = J TO R
190   IF A(I,J) < > 0 THEN 230
200   NEXT I
210   PRINT "NO UNIQUE SOLUTION"
220   GOTO 440
230   FOR K = 1 TO R + 1
240  X = A(J,K)
250  A(J,K) = A(I,K)
260  A(I,K) = X
270   NEXT K
280  Y = 1 / A(J,J)
290   FOR K = 1 TO R + 1
300  A(J,K) = Y * A(J,K)
310   NEXT K
320   FOR I = 1 TO R
330   IF I = J THEN 380
340  Y =  - A(I,J)
350   FOR K = 1 TO R + 1
360  A(I,K) = A(I,K) + Y * A(J,K)
370   NEXT K
380   NEXT I
390   NEXT J
400   PRINT
410   FOR I = 1 TO R
420   PRINT "X";I;"=";  INT (A(I,R + 1) * 1000 + .5) / 1000
430   NEXT I
440   END
```

Compact Crout Method

This program will compute solutions to the system of linear equations:

$$AX = B$$

The program uses a triangular factorization of A in the form $A = L D L^T$ (a $\triangle$ low triangular matrix, D diagonal matrix) implemented according to the Crout-Dollite method.

Since, for reasons of speed, there is no pivotal strategy, the matrix A must be positive definite, as in those systems of equations used in matrix structural analysis, which are generally positive definite.

Additionally, the user need only input the upper triangle of matrix A in a one-dimensional array since only the upper triangle is stored in this program.

Program Notes

When inputting the upper triangle of matrix A, be sure to enter the data in columns.

If $N > 15$, in a matrix which has the dimensions $N \times N$, the dimension statement in line 1 must be changed so that arrays A (120) and B (15) are redimensioned according to the following scheme:

$$A \longrightarrow \text{to } N \times (N + 1)/2$$
$$B \longrightarrow \text{to } N$$

Example

Given:

$$A = (aij) \text{ to be}$$

$$aij = \frac{1}{i + j - 1}$$

For $N = 4$, we then have (after rounding off the decimals):

$$A = \begin{array}{llll} 1.0 & 0.5 & 0.33333 & 0.25 \\ & 0.33333 & 0.25 & 0.2 \\ & & 0.2 & 0.16667 \\ SYMM & & & 0.14286 \end{array}$$

$$B^T = 0.58333, 0.21667, 0.11666, 0.07381$$

```
RUN
COMPACT CROUT METHOD



>ENTER MATRIX A DIMENSION

  N = ?4


>ENTER MATRIX A UPPER TRIANGLE
```

```
  UPPER PART COLUMN 1

  A(1,1) = ?1

    DATA CORRECT (Y/N)?Y


  UPPER PART COLUMN 2

  A(1,2) = ?.5
  A(2,2) = ?.33333

    DATA CORRECT (Y/N)?Y


  UPPER PART COLUMN 3

  A(1,3) = ?.33333
  A(2,3) = ?.25
  A(3,3) = ?.2

    DATA CORRECT (Y/N)?Y


  UPPER PART COLUMN 4

  A(1,4) = ?.25
  A(2,4) = ?.2
  A(3,4) = ?.16667
  A(4,4) = ?.14286

    DATA CORRECT (Y/N)?Y



>DETERMINANT OF A

  DET = 1.61111195E-07



>INPUT VECTOR B

  B(1) = ?.58333
  B(2) = ?.21667
  B(3) = ?.11666
  B(4) = ?.07381


>SOLUTION
 ********

  X(1) = .999999946
  X(2) = -.9999994
  X(3) = .999998563
  X(4) = -.999999068
```

```
 DO YOU WANT TO INPUT ANOTHER
  VECTOR B ? (Y/N) N

>END OF EXECUTION
```

Program Listing

```
1   DIM A(120),B(15)
5   HOME
10   PRINT "COMPACT CROUT METHOD"
20   PRINT : PRINT
170   PRINT
175   REM ==CONTROL PROGRAM==
180 E5 = 1E - 8
185   REM ==READ MATRIX A==
190   GOSUB 1100
195   REM ==FACTORIZATION OF A ==
200   GOSUB 400
210   PRINT
220   PRINT ">DETERMINANT OF A"
230   PRINT
240   PRINT "  DET = ";D8
250   PRINT
260   PRINT
265   REM ==READ VECTOR B==
270   GOSUB 1400
275   REM ==SOLUTION OF A*X=B==
280   GOSUB 800
285   REM ==OUTPUT VECTOR X==
290   GOSUB 1500
300   PRINT
310   PRINT " DO YOU WANT TO INPUT ANOTHER"
320   INPUT "  VECTOR B ? (Y/N) ";W$
330   IF W$ = "N" THEN 360
340   PRINT
350   GOTO 270
360   PRINT
370   PRINT ">END OF EXECUTION"
380   END
395   REM ==SUBROUTINE CROUT==
400 J0 = 1
405   REM ==CHECK FOR ZERO PIVOT==
410 D8 = A(J0)
420   IF  ABS (D8) < E5 THEN 730
425   REM ==LOOP OVER J COLUMNS==
430   FOR J = 2 TO N
440 J2 = J - 1
450   IF J = 2 THEN 570
460 I0 = 1
465   REM ==LOOP OVER I ROWS==
470   FOR I = 2 TO J2
480 I2 = I - 1
490 I1 = J0 + I
495   REM  ==REDUCE ELEMENT A(I,J)==
```

```
500    FOR K = 1 TO I2
510  K1 = I0 + K
520  K2 = J0 + K
530  A(I1) = A(I1) - A(K1) * A(K2)
540    NEXT K
550  I0 = I0 + I
560    NEXT I
565    REM ==REDUCE COLUMN J==
566    REM  ==REDUCE ELEMENT A(J,J)==
570  J3 = J0 + J
580  I3 = 0
590  D1 = A(J3)
600    FOR I = 1 TO J2
610  I3 = I3 + I
620  I1 = J0 + I
630  U = A(I1) / A(I3)
640  D1 = D1 - U * A(I1)
650  A(I1) = U
660    NEXT I
670  A(J3) = D1
675    REM ==CHECK FOR ZERO PIVOT==
680    IF  ABS (D1) < E5 THEN 730
690  J0 = J0 + J
700  D8 = D8 * D1
710    NEXT J
720    RETURN
730    PRINT "   **FATAL ERROR** MATRIX A"
740    PRINT "   IS NOT POSITIVE-DEFINITE"
750    PRINT
760    PRINT
770    END
795    REM ==SUBROUTINE SOLUTION==
796    REM ==FORWARD SOLUTION==
800  J0 = 1
805    REM ===LOOP OVER J COLUMNS==
810    FOR J = 2 TO N
820  J2 = J - 1
825    REM  ==REDUCE COMPONENT B(J)==
830    FOR K = 1 TO J2
840  J4 = J0 + K
850  B(J) = B(J) - A(J4) * B(K)
860    NEXT K
870  J0 = J0 + J
880    NEXT J
886    REM  ==BACK SUBSTITUTION==
890  J0 = 0
895    REM  ==PERFORM B(J)/A(J,J)==
900    FOR J = 1 TO N
910  J0 = J0 + J
920  B(J) = B(J) / A(J0)
930    NEXT J
935    REM  ==BACKWARD' LOOP OVER J COLUMNS==
940  J = N
950  J0 = N * (N + 1) / 2
960    IF (J < = 1) THEN 1050
970  J2 = J - 1
```

```
975   REM   ==SUBSTRACT COLUMNS==
980  J0 = J0 - J
990   FOR I = 1 TO J2
1000  I1 = J0 + I
1010  B(I) = B(I) - A(I1) * B(J)
1020   NEXT I
1030  J = J2
1040   GOTO 960
1050   RETURN
1095   REM   ==INPUT MODULE FOR A==
1100   PRINT ">ENTER MATRIX A DIMENSION"
1110   PRINT
1120   INPUT "   N = ?";N
1130  J0 = 0
1140   PRINT
1150   PRINT
1160   PRINT ">ENTER MATRIX A UPPER TRIANGLE"
1170   FOR J = 1 TO N
1180   PRINT
1190   PRINT
1200   PRINT "   UPPER PART COLUMN ";J
1210   PRINT
1220   FOR I = 1 TO J
1230  I1 = I + J0
1240   PRINT "   A(";I;",";J;") = ";
1250   INPUT A(I1)
1260   NEXT I
1270   PRINT
1280   INPUT "    DATA CORRECT (Y/N)?";W$
1290   IF W$ = "N" THEN 1180
1300  J0 = J0 + J
1310   NEXT J
1320   PRINT
1330   PRINT
1340   RETURN
1395   REM  ==INPUT MODULE FOR B==
1400   PRINT
1410   PRINT ">INPUT VECTOR B"
1420   PRINT
1430   FOR I = 1 TO N
1440   PRINT "   B(";I;") = ";
1450   INPUT B(I)
1460   NEXT I
1470   PRINT
1480   RETURN
1500   PRINT
1510   PRINT ">SOLUTION"
1520   PRINT " ********"
1530   PRINT
1540   FOR I = 1 TO N
1550   PRINT "   X(";I;") = ";B(I)
1560   NEXT I
1570   RETURN
```

References

Forsythe, G., and Moler. *Computer Solutions to Linear Algebraic Systems.* Prentice-Hall, 1967.

Isaacson, E., and Keller, H. B. *Analysis of Numerical Methods.* John Wiley and Sons, 1966.

General Linear Equation Solver

This is a general program, concerned with the evaluation of the system of linear equations:

$$AX = b$$

where: A is an $N \times N$ known matrix
          b is an $N \times 1$ known vector
and X is an $N \times 1$ unknown vector

The purpose of this program is twofold:

1. To estimate the condition of the system of equations, generating a number which indicates how good or bad the system of equations is, regarding the possibility of getting a satisfactory solution.
2. To solve the system (the result of the previous step will indicate the accuracy of our expected solution).

Program Notes

The maximum problem size is currently set to 10 in the DIM statement (line 130). This can easily be modified as follows:

$$130 \ \text{DIM} \ A(N,N), \ B(N)$$

where N is the maximum dimension.

The solution is based on a factorization of $P \times A = L \times U$ where:

       P is a permutation matrix
       L is the lower triangular matrix
       U is the upper triangular matrix

Example

Solve for the system of equations:

$$2X_1 + 3X_2 - X_3 = 5$$
$$4X_1 + 4X_2 - 3X_3 = 3$$
$$-2X_1 + 3X_2 - X_3 = 1$$

This can be represented in matrix notation as:

$$\begin{bmatrix} 2 & 3 & -1 \\ 4 & 4 & -3 \\ 4 & 3 & -1 \end{bmatrix} \begin{bmatrix} X_1 \\ X_2 \\ X_3 \end{bmatrix} = \begin{bmatrix} 5 \\ 3 \\ 1 \end{bmatrix}$$

$$A \quad\quad \times \quad\quad X \quad = \quad B$$

```
RUN
GENERAL LINEAR EQUATION SOLVER



ENTER SIZE OF A(N,N)

 N = 3

INPUT ELEMENTS OF A
COLUMN BY COLUMN


ENTER COLUMN 1

A(1,1) = ?2
A(2,1) = ?4
A(3,1) = ?-2


ENTER COLUMN 2

A(1,2) = ?3
A(2,2) = ?4
A(3,2) = ?3


ENTER COLUMN 3

A(1,3) = ?-1
A(2,3) = ?-3
A(3,3) = ?-1



WANT DET(A) (Y/N) ?Y

DETERMINANT OF A = 20



ESTIMATION OF CONDITION

OF MATRIX A

 C = 16.6578947


INPUT VECTOR B

B(1) = ?5
B(2) = ?3
B(3) = ?1
```

SOLUTION
****

X(1) = 1
X(2) = 2
X(3) = 3

WANT ANOTHER B (Y/N) ?N

Program Listing

```
5   HOME
10   PRINT "GENERAL LINEAR EQUATION SOLVER"
20   PRINT : PRINT
100   REM  ==FACTORIZATION OF A==
130   DIM A(10,10),B(10)
150   REM  ==INPUT DATA==
160   GOSUB 420
180   REM  ==1 - NORM OF A==
190   GOSUB 860
200   REM  ==FACTORIZATION OF A ==
220   GOSUB 1050
230   REM  ==ESTIMATION OF CONDITION OF MATRIX A==
250   GOSUB 1640
270   REM  ==INPUT VECTOR B==
280   GOSUB 650
290   REM ==SOLVE A*X=B ==
310   GOSUB 2440
320   REM
330   REM ==OUTPUT SOLUTION==
340   GOSUB 760
350   PRINT
360   PRINT "WANT ANOTHER B (Y/N) ";
370   INPUT W$
380   IF W$ = "N" THEN 2680
390   GOTO 280
400   REM ==INPUT DATA ==
420   PRINT
430   PRINT
440   PRINT "ENTER SIZE OF A(N,N)"
450   PRINT
460   INPUT " N = ";N
470   PRINT
480   PRINT "INPUT ELEMENTS OF A"
490   PRINT "COLUMN BY COLUMN"
500   PRINT
510   FOR J = 1 TO N
520   PRINT
530   PRINT "ENTER COLUMN ";J
540'  PRINT
550   FOR I = 1 TO N
560   PRINT "A(";I;",";J;") = ";
570   INPUT A(I,J)
580   NEXT I
```

```
590    PRINT
600    NEXT J
610    PRINT
620    RETURN
630    REM ==INPUT B ROUTINE==
650    PRINT
660    PRINT "INPUT VECTOR B"
670    PRINT
680    FOR I = 1 TO N
690    PRINT "B(";I;") = ";
700    INPUT B(I)
710    NEXT I
720    PRINT
730    RETURN
740    REM ==OUTPUT SOLUTION==
760    PRINT
770    PRINT "SOLUTION"
780    PRINT "********"
790    PRINT
800    FOR I = 1 TO N
810    PRINT "X(";I;") = ";B(I)
820    NEXT I
830    RETURN
840    REM == 1 NORM OF MATRIX A ==
860 A9 = 0
870    FOR J = 1 TO N
880 T = 0
890    FOR I = 1 TO N
900 T = T +  ABS (A(I,J))
910    NEXT I
920    IF T <  = A9 THEN 940
930 A9 = T
940    NEXT J
950    RETURN
1000   REM == P IS PERMUTATION MATRIX==
1010   REM == L IS LOWER TRIANGULAR MATRIX===
1020   REM == U IS UPPER TRIANGULAR MATRIX==
1050 IO(N) = 1
1060 N1 = N - 1
1070   IF N1 = 0 THEN 2150
1080   FOR K = 1 TO N1
1090 K1 = K + 1
1100   REM ==FIND PIVOT==
1110 M = K
1120   FOR I = K1 TO N
1130 A1 =   ABS (A(I,K))
1140 A2 =   ABS (A(M,K))
1150   IF A1 <  = A2 THEN 1170
1160 M = I
1170   NEXT I
1180   REM ==PIVOT LOCATED==
1200 IO(K) = M
1210   IF M = K THEN 1250
1230 IO(N) =  - IO(N)
1240   REM ==INTERCHANGE DIAGONAL TERMS==
1250 T = A(M,K)
```

```
1260 A(M,K) = A(K,K)
1270 A(K,K) = T
1280   REM ==SKIP STEP IF PIVOT IS ZERO==
1290   IF T = 0 THEN 1450
1300   REM ==COMPUTE MULTIPLIERS==
1310   FOR I = K1 TO N
1320 A(I,K) =   - A(I,K) / T
1330   NEXT I
1340   REM   ==INTERCHANGE AND ELIMINATE BY COLUMNS==
1360   FOR J = K1 TO N
1370 T = A(M,J)
1380 A(M,J) = A(K,J)
1390 A(K,J) = T
1400   IF T = 0 THEN 1440
1410   FOR I = K1 TO N
1420 A(I,J) = A(I,J) + A(I,K) * T
1430   NEXT I
1440   NEXT J
1450   NEXT K
1460   PRINT
1470   PRINT "WANT DET(A) (Y/N) ";
1480   INPUT W$
1490   IF W$ = "N" THEN 1590
1500   REM   ==COMPUTE DETERMINANT==
1510 D = 1
1520   FOR I = 1 TO N
1530 D = D * A(I,I)
1540   NEXT I
1550 D = D * IO(N)
1560   PRINT
1570   PRINT "DETERMINANT OF A = ";D
1580   PRINT
1590   RETURN
1600   REM    ==ESTIMATION OF CONDITION OF MATRIX A==
1640   IF N > 1 THEN 1740
1650 C = 1
1660   IF A(1,1) <  > 0 THEN 2240
1670   GOTO 2380
1680   REM ==GENERAL CASE==
1740 K = 1
1750 T = 0
1760   IF K = 1 THEN 1810
1770 K1 = K - 1
1780   FOR I = 1 TO K1
1790 T = T + A(I,K) * B(I)
1800   NEXT I
1810 E1 = 1
1820   IF T >  = 0 THEN 1850
1830 E1 =   - 1
1840   REM ==CHECK FOR SINGULARITY==
1850   IF A(K,K) = 0 THEN 2380
1860 B(K) =   - (E1 + T) / A(K,K)
1870 K = K + 1
1880   IF K >  = (N + 1) THEN 1910
1890   GOTO 1750
1900   REM ==SOLVE FOR SYSTEM=
```

```
1910   FOR K2 = 1 TO N1
1920 K = N - K2
1930 T = 0
1940 K1 = K + 1
1950   FOR I = K1 TO N
1960 T = T + A(I,K) * B(K)
1970   NEXT I
1980 B(K) = T
1990 M = IO(K)
2000   IF M = K THEN 2040
2010 T = B(M)
2020 B(M) = B(K)
2030 B(K) = T
2040   NEXT K2
2060 Y1 = 0
2070   FOR I = 1 TO N
2080 Y1 = Y1 +  ABS (B(I))
2090   NEXT I
2110   GOSUB 2440
2130 Z1 = 0
2140   FOR I = 1 TO N
2150 Z1 = Z1 +  ABS (B(I))
2160   NEXT I
2210 C = A9 * Z1 / Y1
2220   IF C >  = 1 THEN 2240
2230 C = 1
2240   PRINT
2250   PRINT
2260   PRINT "ESTIMATION OF CONDITION"
2270   PRINT
2280   PRINT "OF MATRIX A"
2290   PRINT
2300   PRINT " C = ";C
2310   REM ==TEST FOR SINGULARITY==
2320   IF C < (C + 1) THEN 2350
2330   PRINT
2340   PRINT "A IS SINGULAR TO WORKING PRECISION"
2350   PRINT
2360   RETURN
2370   PRINT
2380   PRINT "MATRIX A IS SINGULAR"
2390   GOTO 2680
2430   REM ==FORWARD SOLUTION==
2440   IF N = 1 THEN 2660
2450 N1 = N - 1
2460   FOR K = 1 TO N1
2470 K1 = K + 1
2480 M = IO(K)
2490 T = B(M)
2500 B(M) = B(K)
2510 B(K) = T
2520   FOR I = K1 TO N
2530 B(I) = B(I) + A(I,K) * T
2540   NEXT I
2550   NEXT K
2570   FOR K2 = 1 TO N1
```

```
2580 K1 = N - K2
2590 K = K1 + 1
2600 B(K) = B(K) / A(K,K)
2610 T =   - B(K)
2620  FOR I = 1 TO K1
2630 B(I) = B(I) + A(I,K) * T
2640  NEXT I
2650  NEXT K2
2660 B(1) = B(1) / A(1,1)
2670  RETURN
2680  END
```

References

Forsythe, G., and Moler. *Computer Solutions of Linear Algebraic Systems.* Prentice-Hall, 1967.

Isaacson, E., and Keller, H. B. *Analysis of Numerical Methods.* John Wiley and Sons, 1966.

6
Eigenvalues and Eigenvectors

Inverse Iteration with Shifts

Cyclic Jacobi Method

Eigenvalues of a General Matrix

Inverse Iteration with Shifts

This program uses the inverse iteration method to solve the eigenvalue problem:

$$Ax = \lambda Bx$$

This method converges to the lowest eigenpair (λ_1, E_1), provided the starting iteration vector (X_0) is not orthogonal to E_1.

To avoid the limitations of convergence *only* to the lowest eigenpair, the program allows the user to "shift" to any eigenpair, as shown in the figure below:

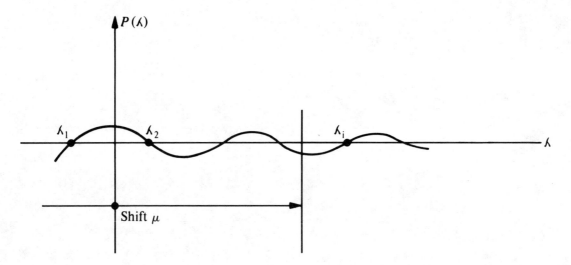

The introduced shift μ will determine convergence to λ (which is the closest value to μ). Furthermore, the introduction of the shift allows the following:

- An improvement in the order of convergence.
- Matrix A does not have to be either positive definite or non-singular.

Program Notes

The data that this program requires is as follows:

- The dimension N.
- The matrix A (Note: since the matrix is assumed to be symmetrical, you need only enter the upper triangle by columns.)
- The matrix B. Two options are available:

    Type 1: B is a diagonal matrix and the user need only enter the diagonal elements.

    Type 2: B is a full symmetrical matrix. Again, the user need only enter the upper triangle.

- Error tolerance E. The solution will only be evaluated for an eigenvalue within the specified tolerance.

The following parameters are under your control during program execution:

- Maximum number of iterations, NMAX, to be performed in the inverse iteration procedure
- Shift

The arrays in this program — A, A_1, B, U, and Z — have been dimensioned for a matrix which is 15×15. For a matrix larger than 15×15, change the dimension statement in line 1 as follows:

```
1 DIM A(N(N+1)/2), B(N(N+1)/2), A1(N(N+1)/2), U(N), Z(N)
```

where N is the maximum size dimension.

The starting iteration vector has been chosen so that its components are the diagonal elements of matrix A divided by the corresponding diagonal elements of matrix B. To redefine this, delete lines 1260 to 1330 and substitute the desired expression.

Example

Consider the following:

$$A = \begin{bmatrix} 0 & 1 & 1 \\ 1 & 0 & 1 \\ 1 & 1 & 0 \end{bmatrix}, \quad B = \begin{bmatrix} 1 & 0 & 0 \\ 0 & 1 & 0 \\ 0 & 0 & 1 \end{bmatrix}$$

```
RUN


INVERSE ITERATION WITH SHIFTS


      TYPE (1) :    DIAGONAL MATRIX
      TYPE (2) :    FULL SYMMETRICAL MATRIX


>ENTER TYPE OF PROBLEM

  TYPE = 1


>DIMENSION MATRIX A

  N = ?3


>ENTER MATRIX A UPPER TRIANGLE

  UPPER PART COLUMN 1

  A(1,1) = ?0


  UPPER PART COLUMN 2

  A(1,2) = ?1
  A(2,2) = ?0
```

```
     UPPER PART COLUMN 3

     A(1,3) = ?1
     A(2,3) = ?1
     A(3,3) = ?0



>INPUT VECTOR B

     B(1) = ?1
     B(2) = ?1
     B(3) = ?1



>ENTER TOLERANCE

     TOL = 1E-8


>MAXIMUM NUMBER OF ITERATIONS

     NMAX = 30

>INPUT DESIRED SHIFT

     SHIFT = 0

               -RUNNING-

     **ERROR** UNSUCCESSFUL FACTORIZATION.
     CHOOSE NEW SHIFT


>INPUT DESIRED SHIFT

     SHIFT = 4

                -RUNNING-


     NUMBER OF EIGENVALUES

     SMALLER THAN  SHIFT = 3


     DETERMINANT = -50


     WANT SHIFT CHANGED (Y/N) ?N

                -RUNNING-
```

```
>SOLUTION
 ********

EIGENVALUE

   EIG = 2

EIGENVECTOR

   U(1)= .577362783
   U(2)= .577352541
   U(3)= .577335472

NUM. ITERATIONS

   N = 9


>WANT ANOTHER EIGENSOLUTION (Y/N) ?Y


>MAXIMUM NUMBER OF ITERATIONS

   NMAX = 20

>INPUT DESIRED SHIFT

   SHIFT = -5

             -RUNNING-


   NUMBER OF EIGENVALUES

   SMALLER THAN  SHIFT = 0


   DETERMINANT = 112


WANT SHIFT CHANGED (Y/N) ?N

           -RUNNING-


**WARNING** CONVERGENCE NOT
ATTAINED. CHANGE NMAX OR SHIFT


WANT SHIFT CHANGED (Y/N) ?N

           -RUNNING-


>MAXIMUM NUMBER OF ITERATIONS
```

```
   NMAX = 30

>SOLUTION
 ********

EIGENVALUE

  EIG = -.999999989

EIGENVECTOR

  U(1)= .749298501
  U(2)= -.0936287234
  U(3)= -.655580197

NUM. ITERATIONS

  N = 24


>WANT ANOTHER EIGENSOLUTION (Y/N) ?Y


>MAXIMUM NUMBER OF ITERATIONS

  NMAX = 20

>INPUT DESIRED SHIFT

  SHIFT = -1.0001

          -RUNNING-


  NUMBER OF EIGENVALUES

  SMALLER THAN  SHIFT = 0


  DETERMINANT = 3.00008955E-08


  WANT SHIFT CHANGED (Y/N) ?N

          -RUNNING-


>SOLUTION
 ********

EIGENVALUE

  EIG = -1

EIGENVECTOR
```

```
   U(1)=  .81649658
   U(2)= -.408217674
   U(3)= -.408278906
```

NUM. ITERATIONS

   N = 3

>WANT ANOTHER EIGENSOLUTION (Y/N) ?Y

>MAXIMUM NUMBER OF ITERATIONS

   NMAX = 20

>INPUT DESIRED SHIFT

   SHIFT = -1

              -RUNNING-

   ERROR UNSUCCESSFUL FACTORIZATION.
   CHOOSE NEW SHIFT

>INPUT DESIRED SHIFT

   SHIFT = -.999999

                 -RUNNING-

   NUMBER OF EIGENVALUES

   SMALLER THAN SHIFT = 2

   DETERMINANT = 3.00143984E-12

   WANT SHIFT CHANGED (Y/N) ?N

                 -RUNNING-

>SOLUTION
 ****

EIGENVALUE

   EIG = -1

EIGENVECTOR

```
   U(1)= -.81649658
   U(2)= .408248596
   U(3)= .408247984

NUM. ITERATIONS

   N = 3


>WANT ANOTHER EIGENSOLUTION (Y/N) ?N
```

Program Listing

```
1    DIM A(120),B(120),A1(120),U(15),Z(15)
5    HOME
30   PRINT "INVERSE ITERATION WITH SHIFTS"
70   PRINT
80   PRINT
100   PRINT
110   PRINT "     TYPE (1) :    DIAGONAL MATRIX"
120   PRINT "     TYPE (2) :    FULL SYMMETRICAL MATRIX"
130   PRINT
150   PRINT
155   REM ==INPUT DATA==
160   PRINT
170 E5 = 1E - 8
180   PRINT ">ENTER TYPE OF PROBLEM"
190   PRINT
200   INPUT "  TYPE = ";I9
210   PRINT
220   PRINT
230   PRINT ">DIMENSION MATRIX A
240   PRINT
250   INPUT "  N = ?";N
260 J0 = 0
270   PRINT
280   PRINT
290   PRINT ">ENTER MATRIX A UPPER TRIANGLE"
300   FOR J = 1 TO N
310   PRINT
320   PRINT "  UPPER PART COLUMN ";J
330   PRINT
340   FOR I = 1 TO J
350 I1 = I + J0
360   PRINT "  A(";I;",";J;") = ";
370   INPUT A1(I1)
380   NEXT I
390   PRINT
420 J0 = J0 + J
430   NEXT J
440   PRINT
450   PRINT
```

```
460    PRINT
465    REM ==INPUT MATRIX OR VECTOR B==
470    ON I9 GOSUB 2450,2610
480    PRINT
490    PRINT
500    PRINT ">ENTER TOLERANCE"
510    PRINT
520    INPUT "   TOL = ";TO
530 I8 = 0
540    PRINT
550    PRINT
560    PRINT ">MAXIMUM NUMBER OF ITERATIONS"
570    PRINT
580    INPUT "   NMAX = ";N8
590    IF I8 = 2 THEN 970
600    PRINT
610    PRINT ">INPUT DESIRED SHIFT"
620    PRINT
630    INPUT "   SHIFT = ";WO
640    PRINT
641    HOME : PRINT "              -RUNNING-"
642    PRINT
645    REM ==INITIALIZE MATRIX A==
650 JJ = 0
660    FOR J = 1 TO N
670    FOR I = 1 TO J
680 IJ = JJ + I
690 A(IJ) = A1(IJ)
700    NEXT I
710 JJ = JJ + J
720    NEXT J
725    REM ==COMPUTE A - SHIFT B==
730    IF I9 = 1 THEN  GOSUB 2550
735    IF I9 = 2 THEN  GOSUB 2780
740    GOSUB 1740
745    REM ==CHECK POLARITY==
750    IF I8 = 1 THEN 600
760 MO = 0
770 II = 0
775    REM ==CHECK STURM SEQUENCE==
780    FOR I = 1 TO N
790 II = II + I
800    IF A(II) > 0 THEN 820
810 MO = MO + 1
820    NEXT I
830    PRINT
840    PRINT "   NUMBER OF EIGENVALUES"
850    PRINT
860    PRINT "   SMALLER THAN   SHIFT = ";MO
870    PRINT
880    PRINT
890    PRINT "   DETERMINANT = ";D8
900    PRINT
910    PRINT
920    PRINT "   WANT SHIFT CHANGED (Y/N) ";
```

```
930    INPUT W$
940    PRINT
945    HOME : PRINT "                -RUNNING-"
946    PRINT
950    IF W$ = "Y" THEN 600
960    IF I8 = 2 THEN 540
965    REM ==PERFORM INVERSE ITERATIONS==
970    GOSUB 1250
975    REM ==CHECK FOR CONVERGENCY==
980    IF I8 = 2 THEN 900
985    REM ==UNDO SHIFT==
990 EO = EO + WO
1000   GOSUB 1060
1010   PRINT
1020   PRINT ">WANT ANOTHER EIGENSOLUTION (Y/N) ";
1030   INPUT W$
1040   IF W$ = "Y" GOTO 540
1050   END
1056   REM :        OUTPUT MODULE
1060   PRINT
1070   PRINT ">SOLUTION"
1080   PRINT " ********"
1090   PRINT
1100   PRINT "EIGENVALUE"
1110   PRINT
1120   PRINT "  EIG = ";EO
1130   PRINT
1140   PRINT "EIGENVECTOR"
1150   PRINT
1160   FOR I = 1 TO N
1170   PRINT "  U(";I;")= ";U(I)
1180   NEXT I
1190   PRINT
1200   PRINT "NUM. ITERATIONS"
1210   PRINT
1220   PRINT "  N = ";NO
1230   PRINT
1240   RETURN
1245   REM ==INVERSE ITERATION SUBROUTINE==
1250   I8 = 0
1255   REM ==INITIAL EIGENVECTOR==
1260 II = 0
1270   FOR I = 1 TO N
1280 II = II + I
1290 B1 = B(I)
1300   IF I9 = 1 THEN 1320
1310 B1 = B(II)
1320 U(I) = A(II) / B1
1330   NEXT I
1340 E1 = 1E32
1350 NO = 0
1360 R1 = 0
1370   GOTO 1500
1375   REM ==SET ITERATION COUNTER==
1380 NO = NO + 1
```

```
1390   FOR I = 1 TO N
1400 U(I) = Z(I)
1410   NEXT I
1415   REM ==FORWARD SOLUTION==
1420   GOSUB 2180
1425   REM ==COMPUTE RALEIGH QUOTIENT NUMERATOR==
1430 R1 = 0
1440 II = 0
1450   FOR I = 1 TO N
1460 II = II + I
1470 R1 = R1 + U(I) * U(I) / A(II)
1480   NEXT I
1485   REM ==BACK SUBSTITUTION==
1490   GOSUB 2280
1495   REM ==COMPUTE B-NORM OF U==
1500   IF I9 = 1 THEN   GOSUB 2870
1501   IF I9 = 2 THEN   GOSUB 2910
1510 V2 = 0
1520   FOR I = 1 TO N
1530 V2 = V2 + Z(I) * U(I)
1540   NEXT I
1550 V3 =   SQR (V2)
1555   REM ==SCALE Z BY NORM OF U==
1560   FOR I = 1 TO N
1570 Z(I) = Z(I) / V3
1580   NEXT I
1585   REM ==EIGENVALUE ESTIMATION==
1590 E0 = R1 / V2
1600 H1 =   ABS (E1)
1610 H0 =   ABS (E0)
1615   REM ==TEST OF CONVERGENCE==
1620   IF   ABS (H1 - H0) < T0 * H0 THEN 1700
1630 E1 = E0
1640   IF N0 < N8 THEN 1380
1645   REM ==ERROR MESSAGES==
1650 I8 = 2
1660   PRINT
1670   PRINT "   **WARNING** CONVERGENCE NOT"
1680   PRINT "   ATTAINED. CHANGE NMAX OR SHIFT"
1690   RETURN
1695   REM ==FINAL EIGENVECTOR==
1700   FOR I = 1 TO N
1710 U(I) = U(I) / V3
1720   NEXT I
1730   RETURN
1735   REM ==FACTORIZATION OF A ==
1740 I8 = 0
1750 J0 = 1
1760 D8 = A(J0)
1765   REM : LOOP OVER COLUMNS
1770   FOR J = 2 TO N
1780 J2 = J - 1
1790   IF J = 2 THEN 1910
1800 I0 = 1
1805   REM : LOOP OVER ROWS
```

```
1810   FOR I = 2 TO J2
1820  I2 = I - 1
1830  I1 = J0 + I
1835   REM : REDUCE COLUMN I1
1840   FOR K = 1 TO I2
1850  K1 = I0 + K
1860  K2 = J0 + K
1870  A(I1) = A(I1) - A(K1) * A(K2)
1880   NEXT K
1890  I0 = I0 + I
1900   NEXT I
1905   REM ==CHECK FOR ZERO PIVOT==
1910   IF  ABS (A(J0)) < E5 THEN 2120
1915   REM ==REDUCE DIAGONAL ELEMENT AND COLUMN J==
1920  J3 = J0 + J
1930  I3 = 0
1940  D1 = A(J3)
1950   FOR I = 1 TO J2
1960  I3 = I3 + I
1970  I1 = J0 + I
1980  U = A(I1) / A(I3)
1990  D1 = D1 - U * A(I1)
2000  A(I1) = U
2010   NEXT I
2020  A(J3) = D1
2030  D8 = D8 * D1
2040  J0 = J0 + J
2050   NEXT J
2055   REM ==CHECK FOR ZERO PIVOT==
2060   IF  ABS (A(J0)) > E5 THEN 2160
2065   REM : ERROR MESSAGES
2070   PRINT
2080   PRINT "   **WARNING** ZERO DIAGONAL"
2090   PRINT "   ELEMENT. CHANGE SHIFT"
2100   PRINT
2110   RETURN
2120   PRINT "   **ERROR** UNSUCCESSFUL FAC"
2130   PRINT "   TORIZATION. CHOOSE NEW SHIFT"
2140  I8 = 1
2150   PRINT
2160   RETURN
2174   REM ==FORWARD SOLUTION==
2180  J0 = 1
2190   FOR J = 2 TO N
2200  J2 = J - 1
2210   FOR K = 1 TO J2
2220  J4 = J0 + K
2230  U(J) = U(J) - A(J4) * U(K)
2240   NEXT K
2250  J0 = J0 + J
2260   NEXT J
2270   RETURN
2274   REM ==BACK-SUBSTITUTION==
2280  J0 = 0
2290   FOR J = 1 TO N
```

```
2300  JO = JO + J
2310  U(J) = U(J) / A(JO)
2320   NEXT J
2330  J = N
2340  JO = N * (N + 1) / 2
2350   IF (J <  = 1) THEN 2440
2360  J2 = J - 1
2370  JO = JO - J
2380   FOR I = 1 TO J2
2390  I1 = JO + I
2400  U(I) = U(I) - A(I1) * U(J)
2410   NEXT I
2420  J = J2
2430   GOTO 2350
2440   RETURN
2447   REM ==VECTOR B INPUT==
2450   PRINT ">INPUT VECTOR B"
2460   PRINT
2470   FOR I = 1 TO N
2480   PRINT "  B(";I;") = ";
2490   INPUT B(I)
2500   NEXT I
2510   PRINT
2540   RETURN
2550  JJ = 0
2560   FOR J = 1 TO N
2570  JJ = JJ + J
2580  A(JJ) = A(JJ) - WO * B(J)
2590   NEXT J
2600   RETURN
2605   REM ==FULL MATRIX B INPUT==
2610   PRINT ">ENTER MATRIX B UPPER TRIANGLE"
2620  JO = 0
2630   FOR J = 1 TO N
2640   PRINT
2650   PRINT "  UPPER PART COLUMN ";J
2660   PRINT
2670   FOR I = 1 TO J
2680  I1 = I + JO
2690   PRINT "  B(";I;",";J;") = ";
2700   INPUT B(I1)
2710   NEXT I
2720   PRINT
2730   INPUT "  IS IT OK? (Y/N) ";W$
2740   IF W$ = "N" THEN 2640
2750  JO = JO + J
2760   NEXT J
2770   RETURN
2780  JJ = 0
2790   FOR J = 1 TO N
2800   FOR I = 1 TO J
2810  IJ = JJ + I
2820  A(IJ) = A(IJ) - WO * B(IJ)
2830   NEXT I
2840  JJ = JJ + J
```

```
2850   NEXT J
2860   RETURN
2870   FOR I = 1 TO N
2880  Z(I) = B(I) * U(I)
2890   NEXT I
2900   RETURN
2905   REM ==B IS A FULL MATRIX==
2910  JO = 0
2920  N1 = N - 1
2930   FOR J = 1 TO N1
2940  Z(J) = 0
2950   FOR K = 1 TO J
2960  J1 = JO + K
2970  Z(J) = Z(J) + B(J1) * U(K)
2980   NEXT K
2990  JO = JO + J
3000  KO = J + 1
3010  J1 = JO
3020   FOR K = KO TO N
3030  J1 = J1 + K - 1
3040  Z(J) = Z(J) + B(J1) * U(K)
3050   NEXT K
3060   NEXT J
3070  Z(N) = 0
3080   FOR K = 1 TO N
3090  J1 = JO + K
3100  Z(N) = Z(N) + B(J1) * U(K)
3110   NEXT K
3120   RETURN
```

References

Bathe and Wilson. *Numerical Methods in Finite Element Analysis.* Englewood Cliffs, N.J.: Prentice-Hall, 1976.

Clough and Penzien. *Dynamics of Structures.* New York: McGraw-Hill, 1950.

Wilkinson, J. H. *The Algebraic Eigenvalue Problem.* Oxford: Clarendon Press, 1961.

Cyclic Jacobi Method

This program uses the Cyclic Jacobi method to solve the symmetric eigenvalue problem:

$$Ax = \lambda Bx$$

Program Notes

The inputs to this program are as follows:

- N, the size of the problem to be solved
- The number of significant figures desired in the eigenvalues
- The elements of matrix A (only the upper triangle need be entered)
- The elements of matrix B (only the diagonal need be entered)

The maximum problem size allowed in this program is controlled by the DIM statement at line 10. To alter the problem size, modify line 10 as follows:

```
10 DIM A(N,N), B(N), U(N,N)
```

where N is the maximum size dimension.

The eigensolution is normalized with respect to matrices A and B as follows:

$$\begin{aligned} X^T Bx &= I \\ X^T Ax &= \lambda \end{aligned}$$

and where I is the identify matrix

Example

Compute the natural vibration frequencies and mode shapes of the shear building shown below:

$M_1 = 1.0\,K - \sec^2/\text{in.}$
$M_2 = 1.5$
$M_3 = 2.0$
$K_1 = 600\text{k/in.}$
$K_2 = 1200$
$K_2 = 1800$

The free vibration dynamic equilibrium equation can be reduced to:

$$K0 = W^2 M0,$$

where K is the stiffness matrix
M is the mass matrix of the structure
W is the natural circular frequency
0 is the corresponding vibration mode shape

In this case:

$$K = \begin{bmatrix} k_1 & -k_1 & \\ -k_1 & (k_1 + k_2) & -k \\ & -k_2 & k_2 + k_3 \end{bmatrix} = \begin{bmatrix} 600 & -600 & 0 \\ -600 & 1800 & -1200 \\ 0 & -1200 & 3000 \end{bmatrix}$$

$$M = \begin{bmatrix} m_1 & 0 & 0 \\ 0 & m_2 & 0 \\ 0 & 0 & m_3 \end{bmatrix} = \begin{bmatrix} 1.0 & 0 & 0 \\ 0 & 1.5 & 0 \\ 0 & 0 & 2.0 \end{bmatrix}$$

```
RUN

CYCLIC JACOBI METHOD


ENTER SIZE OF PROBLEM

N= ? 3

NUMBER OF SIGNIFICANT FIGURES? 5


ENTER UPPER TRIANGLE

OF MATRIX A , BY COLUMNS




ENTER UPPER PART OF COLUMN 1

A(1,1) = ?600




ENTER UPPER PART OF COLUMN 2

A(1,2) = ?-600
A(2,2) = ?1800




ENTER UPPER PART OF COLUMN 3

A(1,3) = ?0
A(2,3) = ?-1200
A(3,3) = ?3000




ENTER ELEMENTS OF
```

DIAGONAL MATRIX B

B(1) = ?1.0
B(2) = ?1.0
B(3) = ?2.0

 TOLERANCE= 9.99999998E-11

SOLUTION
****

NO. OF ROTATIONS REQUIRED = 7

EIGENVALUE 1 IS 237.285068

 ITS EIGENVECTOR IS

.808326479
.488653471
.232191824

PRESS THE SPACE BAR TO CONTINUE

EIGENVALUE 2 IS 1042.06808

 ITS EIGENVECTOR IS

-.54179432
.399182796
.523025815

PRESS THE SPACE BAR TO CONTINUE

EIGENVALUE 3 IS 2620.64684

 ITS EIGENVECTOR IS

.230363225
-.775803375
.415368454

PRESS THE SPACE BAR TO CONTINUE

END OF PROGRAM

Program Listing

```
4    HOME
5    REM ==CYCLIC JACOBI METHOD==
10   DIM A(10,10),B(10),U(10,10)
19   REM ==INPUT DATA==
20   GOSUB 70
28   REM ==INITIALIZE==
30   GOSUB 660
40   GOSUB 980
46   REM ==OUTPUT RESULTS==
50   GOSUB 750
60   GOTO 3000
65   REM ==INPUT DATA==
70   PRINT
80   PRINT " CYCLIC JACOBI METHOD"
160  PRINT
180  PRINT
190  PRINT "ENTER SIZE OF PROBLEM"
200  PRINT
210  INPUT "N= ? ";N
220  PRINT
230  INPUT "NUMBER OF SIGNIFICANT FIGURES? ";S1
240  PRINT
250  PRINT
260  PRINT "ENTER UPPER TRIANGLE "
270  PRINT
280  PRINT "OF MATRIX A , BY COLUMNS"
290  PRINT
300  PRINT
310  PRINT
320  FOR J = 1 TO N
330  PRINT
340  PRINT "ENTER UPPER PART OF COLUMN ";J
350  PRINT
360  FOR I = 1 TO J
370  PRINT "A(";I;",";J;") = ";
380  INPUT A(I,J)
390  NEXT I
400  PRINT
440  PRINT
450  NEXT J
460  FOR I = 1 TO N
470  I1 = I - 1
480  FOR J = 1 TO I1
490  A(I,J) = A(J,I)
500  NEXT J
510  NEXT I
520  PRINT
530  PRINT "ENTER ELEMENTS OF "
540  PRINT
550  PRINT "DIAGONAL MATRIX B"
560  PRINT
570  FOR I = 1 TO N
580  PRINT "B(";I;") = ";
```

```
590    INPUT B(I)
600    NEXT I
610    PRINT
650    RETURN
655    REM ==INITIALIZE TOLERANCE AND MAX NO. OF ROTATIONS==
660 Z = 2 * S1
670 T1 = 1 / (10 ^ Z)
680    PRINT
690    HOME : PRINT " TOLERANCE= ";T1
700 R = 5 * N * N
710 R1 = 0
720 T2 = 0.1
730 N1 = N - 1
740    RETURN
746    REM ==OUTPUT EIGENSOLUTION==
750    PRINT
760    PRINT
770    PRINT "SOLUTION"
780    PRINT "********"
790    PRINT
800    PRINT "NO. OF ROTATIONS REQUIRED = ";R1
810    PRINT
820    PRINT
830    FOR J = 1 TO N
840    PRINT
850    PRINT "EIGENVALUE ";J;" IS ";B(J)
860    PRINT
870    PRINT " ITS EIGENVECTOR IS "
880    PRINT
890    FOR I = 1 TO N
900    PRINT U(I,J)
910    NEXT I
920    PRINT
930    PRINT "PRESS THE SPACE BAR TO CONTINUE"
940    GET C$: HOME
950    PRINT
960    NEXT J
970    RETURN
975    REM ==EIGENPROBLEM SOLUTION==
980    GOSUB 1130
985    REM ==PERFORM ONE CYCLE OF ROTATIONS==
990    GOSUB 1290
995    REM ==CHECK TOLERANCE==
1000   IF X1 < T1 THEN 1110
1005   REM  CHECK NO. OF ROTATIONS
1010   IF R1 > R THEN 1040
1020 T2 = 0.1 * X1
1030   GOTO 990
1040   PRINT
1050   PRINT " *** ERROR ***"
1060   PRINT
1070   PRINT "NO CONVERGENCE ATTAINED "
1080   PRINT
1090   PRINT "WITH ";R1;" ROTATIONS"
1100   END
```

```
1110    GOSUB 1830
1120    RETURN
1130    FOR I = 1 TO N
1140    FOR J = 1 TO N
1150 U(I,J) = 0
1160    NEXT J
1170 U(I,I) = 1
1180    NEXT I
1190    FOR I = 1 TO N
1200 B1 =  SQR (B(I))
1210 B(I) = 1 / B1
1220    NEXT I
1230    FOR I = 1 TO N
1240    FOR J = 1 TO N
1250 A(I,J) = B(I) * A(I,J) * B(J)
1260    NEXT J
1270    NEXT I
1280    RETURN
1290 X1 = 0
1300    FOR K = 1 TO N1
1310 K1 = K + 1
1320    FOR L = K1 TO N
1330 A1 = A(K,K)
1340 A2 = A(K,L)
1350 A3 = A(L,L)
1360 X = A2 * A2 / (A1 * A3)
1365    REM ==CHECK IF ROTATION IS NEEDED==
1370    IF X > X1 THEN 1390
1380    GOTO 1400
1390 X1 = X
1400    IF X < T2 THEN 1800
1410 R1 = R1 + 1
1415    REM ==COMPUTE ANGLE==
1420    IF A1 = A3 THEN 1470
1430 Z = 0.5 * (A1 - A3) / A2
1440 Z1 = 1 + 1 / (Z * Z)
1450 T =  - Z * (1 +  SQR (Z1))
1460    GOTO 1480
1470 T = 1
1480 C = 1 /  SQR (1 + T * T)
1490 S = C * T
1500 S2 = S * S
1510 C2 = C * C
1520 A(K,L) = 0
1525    REM ==TRANSFORM DIAGONAL ELEMENTS==
1530 A0 = 2 * A2 * C * S
1540 A(K,K) = A1 * C2 + A0 + A3 * S2
1550 A(L,L) = A1 * S2 - A0 + A3 * C2
1555    REM ==TRANSFORM OFF DIAGONAL ELEMENTS==
1560    FOR I = 1 TO N
1570    IF I < K THEN 1600
1580    IF I > K THEN 1640
1590    GOTO 1740
1600 A0 = A(I,K)
1610 A(I,K) = C * A0 + S * A(I,L)
```

```
1620 A(I,L) =  - S * AO + C * A(I,L)
1630  GOTO 1740
1640  IF I < L THEN 1670
1650  IF I > L THEN 1710
1660  GOTO 1740
1670 AO = A(K,I)
1680 A(K,I) = C * AO + S * A(I,L)
1690 A(I,L) =  - S * AO + C * A(I,L)
1700  GOTO 1740
1710 AO = A(K,I)
1720 A(K,I) = C * AO + S * A(L,I)
1730 A(L,I) =  - S * AO + C * A(L,I)
1740  NEXT I
1745  REM ==TRANSFORM MATRIX U TO GENERATE EIGENVECTORS==
1750  FOR I = 1 TO N
1760 UO = U(I,K)
1770 U(I,K) = C * UO + S * U(I,L)
1780 U(I,L) =  - S * UO + C * U(I,L)
1790  NEXT I
1800  NEXT L
1810  NEXT K
1820  RETURN
1825  REM ==NORMALIZE EIGENVECTORS==
1830  FOR I = 1 TO N
1840  FOR J = 1 TO N
1850 U(I,J) = U(I,J) * B(I)
1860  NEXT J
1870  NEXT I
1880  FOR I = 1 TO N
1890 B(I) = A(I,I)
1900  NEXT I
1905  REM ==ORDER EIGENSOLUTION==
1910  FOR I = 1 TO N1
1920 I1 = I + 1
1930 Z = B(I)
1940 M = I
1950  FOR J = I1 TO N
1960  IF Z < B(J) THEN 1990
1970 Z = B(J)
1980 M = J
1990  NEXT J
2000 B(M) = B(I)
2010 B(I) = Z
2020  FOR J = 1 TO N
2030 Z = U(J,I)
2040 U(J,I) = U(J,M)
2050 U(J,M) = Z
2060  NEXT J
2070  NEXT I
2080  RETURN
3000  PRINT "END OF PROGRAM"
3010  END
```

References

Bathe, Klaus-Jurgen, and Wilson, Edward L. *Numerical Methods and Finite Element Analysis.* Englewood Cliffs, N.J.: Prentice-Hall, 1976.

Clough, Ray W., and Peuzien, Joseph. *Dynamics of Structures.* New York: McGraw-Hill, 1975.

Wilson, E. L. *CAL Computer Analysis Language for the Static and Dynamic Analysis of Structural Systems.* Department of Civil Engineering, University of California, Berkeley, January 1977.

Eigenvalues of a General Matrix

This program computes the eigenvalues of a general matrix (either symmetrical or non-symmetrical). Given:

$$AX = X,$$

we have the determinant:

$$\{A - \lambda 1\}$$

as the characteristic polynomial. If the elements of matrix A are real, then the roots of the polynomial will be real and/or conjugate complex.

The method presented here was developed by Rutishauser, and is quite good for systems up to $N = 30$ or 40 when used with computers accurate to 10-15 digits.

Program Notes

The dimensions of the computer system are limited by the dimension statement at line 40. The system size may be altered as follows:

```
40 DIM A(N,N), L(N,N), U(N,N)
```

where N is the maximum size desired.

Example

Compute the eigenvalues of the following general matrix:

$$A = \begin{bmatrix} -1 & 16 & -20 \\ 1 & 0 & 0 \\ 0 & 1 & 0 \end{bmatrix}$$

Note: The exact values are -5, 2, and 2.

```
RUN
EIGENVALUES OF A GENERAL MATRIX

KEY IN ORDER OF MATRIX 3

INPUT ELEMENTS OF MATRIX A

    A(1,1) =?1
    A(1,2) =?54
    A(1,3) =?-144

    A(2,1) =?1
    A(2,2) =?0
    A(2,3) =?0
```

```
A(3,1) =?0
A(3,2) =?1
A(3,3) =?0



            RUNNING

NUMBER OF CYCLES CONPLETED = 88

ELEMENTS OF UPPER TRIANGULAR MATRIX

                -7.9999999
                5.99999987
                3.00000001
```

Program Listing

```
10   HOME
15 C = 0
20   PRINT "EIGENVALUES OF A GENERAL MATRIX"
30   PRINT
40   DIM A(20,20),L(20,20),U(20,20)
50   INPUT "KEY IN ORDER OF MATRIX ";N
60   PRINT
70   PRINT "INPUT ELEMENTS OF MATRIX A"
80   PRINT
90   FOR I = 1 TO N
95   FOR J = 1 TO N
100   PRINT "    A(";I;",";J;") =";
110   INPUT A(I,J)
120   NEXT J: PRINT : NEXT I
130 M = 0
140   FOR I = 1 TO N: FOR J = 1 TO N
150 L(I,J) = 0:U(I,J) = 0
160   NEXT J: NEXT I
170   FOR K = 1 TO N
180 L(K,K) = 1:U(K,K) = 1
190   NEXT K
191   HOME
192   PRINT : PRINT
193   PRINT "            RUNNING"
200   FOR H = 1 TO N
210 U(1,H) = A(1,H)
220   NEXT H
230   FOR H = 2 TO N
240 L(H,1) = A(H,1) / A(1,1)
250 U(2,H) = A(2,H) - L(2,1) * U(1,H)
260   NEXT H
270 J = 2
280   FOR I = J TO N
290   IF I < = J THEN 350
300 T = 0
310   FOR K = 1 TO J - 1
```

```
320 T = T + L(I,K) * U(K,J)
330   NEXT K
340 L(I,J) = (A(I,J) - T) / U(J,J)
350   NEXT I
360 J = J + 1
370   IF J > N THEN 460
380   FOR I = J TO N
390 T = 0
400   FOR K = 1 TO J - 1
410 T = T + L(J,K) * U(K,I)
420   NEXT K
430 U(J,I) = A(J,I) - T
440   NEXT I
450   GOTO 280
460 M = M + 1
470   IF C = 0 THEN A1 = U(1,1):C = 1: GOTO 520
480 A0 = U(1,1):C = 0
520   FOR I = 1 TO N
530   FOR J = 1 TO N
540 A(I,J) = 0
550   FOR K = 1 TO N
560 A(I,J) = A(I,J) + U(I,K) * L(K,J)
570   NEXT K: NEXT J: NEXT I
580   IF A0 < > A1 THEN   GOTO 200
590   HOME
600   PRINT
610   PRINT "    NUMBER OF CYCLES COMPLETED = ";M
620   PRINT
630   PRINT "   ELEMENTS OF UPPER TRIANGULAR MATRIX"
640   PRINT
660   FOR I = 1 TO N
670   PRINT "                    ";U(I,I)
680   NEXT I
690   END
```

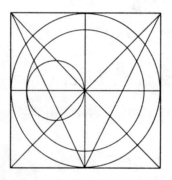

7

Ordinary Differential Equations

Runge-Kutta Method for a First Order Equation

This program computes the initial value for the equation:

$$\left.\begin{array}{l} \dfrac{dYi}{dx} = fi\,(Yj\,(x)\,,x),\ x > a \\ Yi\,(a) = Yai \end{array}\right\} \quad (i = 1,2,\ldots, N)$$

using a 4th order Runge-Kutta method. You will need to enter an end point (b) so that $x \in (a,b)$. You will also need to define the functions $fi\,(\,Yj\,(x),x)$ starting at line 2000. You must also specify a *local* error tolerance (E) and step size (H) for the independent variable (x). Two options are available:

1. Fixed step size. The program checks the local error tolerance at each integration step, prompting an error message if the tolerance is exceeded.
2. Controlled step size. The program automatically controls the step size, increasing it or decreasing it to meet the prescribed tolerance.

Program Notes

As written, this program will solve for systems of the 4th order at most. To increase this, the DIM statement in line 1 must be modified as follows:

```
1 DIM Y(N), YO(N), W(N), Z(N), S(N), Q1(N)
```
where N is the new dimension.

Example

Given the following differential equation:

$$\frac{d^2Y}{dx^2} = 2Y^3$$

with the initial conditions:

$$Y(4) = 1/4$$
$$Y(a) = 1/16$$

the exact solution initially is:

$$Y = 1/x$$

and the proposed differential equation is equivalent to the system:

$$\frac{dY_1}{dx} = Y_2(x)$$

$$dY_2(x) = 2x\,Y_1^3(x)$$

$$Y_1(4) = 0.25 \quad ; \quad Y_2(4) = -0.0625$$

Compute the values.

```
RUN
 RUNGE-KUTTA METHOD FOR A
 FIRST ORDER EQUATION


>NUMBER OF EQUATIONS

    M = 4

>ENTER LIMITS OF INTEGRATION

    A = 0

    B = 1

>ENTER INITIAL DATA

    Y1(A) = ?0

    Y2(A) = ?3

    Y3(A) = ?0

    Y4(A) = ?5

>INITIAL STEPSIZE

    H = .1

>LOCAL ERROR TOLERANCE

    E = 1E-6

>STEPSIZE CONTROL ? (Y/N) ?Y


SOLUTION
********


> X = 0
  Y1 = 0
  Y2 = 3
  Y3 = 0
  Y4 = 5


> X = .1
  Y1 = .297019901
  Y2 = 2.94059602
  Y3 = .446308539
  Y4 = 3.92973829
```

```
> X = .3
  Y1 = .873531997
  Y2 = 2.8252936
  Y3 = 1.0231348
  Y4 = 1.85237304


> X = .5
  Y1 = 1.42743873
  Y2 = 2.71451226
  Y3 = 1.19268778
  Y4 = -.143537554


> X = .7
  Y1 = 1.95962647
  Y2 = 2.60807471
  Y3 = .970936886
  Y4 = -2.06118738


> X = .9
  Y1 = 2.47094683
  Y2 = 2.50581064
  Y3 = .373225369
  Y4 = -3.90364507


> X = 1.1
  Y1 = 2.96221804
  Y2 = 2.4075564
  Y3 = -.585705146
  Y4 = -5.67385897

>END OF EXECUTION
```

Program Listing

```
0    HOME
1    DIM Y(5),Y0(5),W(5),Z(5),S(5),Q1(5)
5    REM ==RUNGE-KUTTA METHOD FOR A FIRST ORDER EQUATION==
7    REM : READ DATA
10    GOSUB 150
20   X0 = A
30   H = H0
40   E5 = E1 / 100
50    PRINT
60    PRINT "> X = ";X0
70    FOR K = 1 TO M
80    PRINT " Y";K;" = ";Y0(K)
90    NEXT K
100    PRINT
110    GOSUB 650
115    REM ==CHECK FOR END INTERVAL==
120    IF (X0 - H) < B THEN 50
130    PRINT ">END OF EXECUTION"
```

```
140    END
150    PRINT " RUNGE-KUTTA METHOD FOR A"
160    PRINT " FIRST ORDER EQUATION"
290    PRINT
300    PRINT
310    PRINT ">NUMBER OF EQUATIONS"
320    PRINT
330    INPUT "      M = ";M
340    PRINT
350    PRINT ">ENTER LIMITS OF INTEGRATION"
360    PRINT
370    INPUT "      A = ";A
380    PRINT
390    INPUT "      B = ";B
400    PRINT
410    PRINT ">ENTER INITIAL DATA"
420    PRINT
430    FOR K = 1 TO M
440    PRINT "     Y";K;"(A) = ";
450    INPUT YO(K)
460    PRINT
470    NEXT K
480    PRINT ">INITIAL STEPSIZE "
490    PRINT
500    INPUT "      H = ";HO
510    PRINT
520    PRINT ">LOCAL ERROR TOLERANCE"
530    PRINT
540    INPUT "      E = ";E1
550    PRINT
560    PRINT ">STEPSIZE CONTROL ? (Y/N) ";
570    INPUT C$
580    PRINT
590    PRINT
600    PRINT "SOLUTION"
610    PRINT "********"
620    PRINT
630    RETURN
650 T2 = H
660 T = T2
670    GOSUB 1100
680    FOR K = 1 TO M
690 Q1(K) = W(K)
700    NEXT K
710 T = T2 / 2
720    GOSUB 1100
730    FOR K = 1 TO M
740 Z(K) = W(K)
750 S(K) = Z(K)
760    NEXT K
770 X = XO + T
780    GOSUB 1140
790 I8 = 0
800    FOR K = 1 TO M
810 Q3(K) = (W(K) - Q1(K)) / 15
```

```
820 D =  ABS (Q3(K)) / T2
830  IF D >  = E1 THEN 960
840  IF D < E5 THEN 860
850 I8 = 1
860  NEXT K
865  REM  ==CONVERGENCE ATTAINED==
870  FOR K = 1 TO M
880 YO(K) = W(K) + Q3(K)
890  NEXT K
900 H = T2
910 XO = XO + H
915  REM ==CHECK FOR STEP-SIZE CONTROL==
920  IF C$ = "N" THEN 950
930  IF I8 = 1 THEN 950
940 H = 2 * H
950  RETURN
955  REM ==CONVERGENCE NOT ATTAINED==
960  IF C$ = "N" THEN 1020
970 T2 = T2 / 2
980  FOR K = 1 TO M
990 Q1(K) = Z(K)
1000  NEXT K
1010  GOTO 710
1020  PRINT
1030  PRINT "  **ERROR** CONVERGENCE"
1040  PRINT "    NOT ATTAINED WITHIN"
1050  PRINT "    REQUIRED TOLERANCE "
1060  PRINT "    FOR GIVEN STEP-SIZE"
1070  PRINT "     H=";H;"  AT X=";XO
1080  END
1085  REM ==RUNGE-KUTTA MODULE==
1087  REM  ==INITIALIZE VARIABLES ==
1100 X = XO
1110  FOR K = 1 TO M
1120 S(K) = YO(K)
1130  NEXT K
1140  FOR K = 1 TO M
1150 Y(K) = S(K)
1160 W(K) = S(K)
1170  NEXT K
1175  REM ==COMPUTE W VALUE==
1180 K = 0
1190 K = K + 1
1200  GOSUB 2000
1210 D = T * F
1220 W(K) = W(K) + D / 6
1230 Y(K) = S(K) + D / 2
1240  IF K < M THEN 1190
1250 X = X + T / 2
1260 K = 0
1270 K = K + 1
1280  GOSUB 2000
1290 D = T * F
1300 W(K) = W(K) + D / 3
```

```
1310 Y(K) = S(K) + D / 2
1320  IF K < M THEN 1270
1330 K = O
1340 K = K + 1
1350  GOSUB 2000
1360 D = T * F
1370 Y(K) = S(K) + D
1380 W(K) = W(K) + D / 3
1390  IF K < M THEN 1340
1400 X = X + T / 2
1410 K = O
1420 K = K + 1
1430  GOSUB 2000
1440 D = T * F
1450 W(K) = W(K) + D / 6
1460  IF K < M THEN 1420
1470  RETURN
2000  IF K = 1 THEN 2010
2001  IF K = 2 THEN 2030
2002  IF K = 3 THEN 2050
2003  IF K = 4 THEN 2070
2004  IF K = 5 THEN 2090
2010 F = Y(2)
2020  RETURN
2030 F =  - .2 * Y(2)
2040  RETURN
2050 F = Y(4)
2060  RETURN
2070 F =  - 9.81 - .2 * Y(4)
2080  RETURN
2090 F =
2100  RETURN
```

References

Bakhualon, N. S. *Numerical Methods.* Moscow: Mir, 1972.

Conte, Samuel D., and De Boor, Carl. *Elementary Numerical Analysis.* McGraw-Hill, 1972.

Dahlquist, Bjorck Anderson. *Numerical Methods.* Prentice-Hall, 1974.

Gear, C. W. *Numerical Initial Value Problems in Ordinary Differential Equations.* Prentice-Hall, 1971.

Predictor-Corrector Method for a First Order Equation

This program computes the solution $Y(X)$ of the ordinary differential equation using the Adams-Baskfort and Adams-Moulton methods:

$$\frac{dy(x)}{dx} = f(x,y)$$

for:

$$a < x < b$$

given the initial condition:

$$y(a) = ya$$

Program Notes

In order to use this program you will need to provide the following information:

- Define the function $f(x,y)$ in line 10 using a DEF FN F(x)= statement.
- Specify the input step size.
- Define a tolerance for the local error.
- Specify the maximum number of correction cycles to be run.

Example

Solve the equation:

$$\frac{dy}{ax} = f(x,y) = x + y$$

in the interval:

$$(0,1)$$

for the initial condition:

$$y(0) = 0$$

```
10 DEF FN F(X) = X + Y

]RUN

PREDICTOR-CORRECTOR METHOD
FOR A FIRST ORDER EQUATION


ENTER INTERVAL OF INTEGRATION

A = 0
B = 1

INITIAL CONDITION
```

```
YA = 0

WANT STEPSIZE CONTROL (Y/N) ?N

INITIAL STEPSIZE

H = 0.25

TOLERANCE OF LOCAL ERROR = 1E-7

MAXIMUM NUMBER OF

CORRECTION CYCLES ?3


SOLUTION
********


  X = .25  Y = .0340169271
  X = .5  Y = .148699469
  X = .75  Y = .366958026

*** ERROR ***

NO CONVERGENCE ATTAINED

WITHIN SPECIFIED TOLERANCE

RUN

PREDICTOR-CORRECTOR METHOD
FOR A FIRST ORDER EQUATION


ENTER INTERVAL OF INTEGRATION

A = 0
B = 1

INITIAL CONDITION

YA = 0

WANT STEPSIZE CONTROL (Y/N) ?Y

INITIAL STEPSIZE

H = .5

TOLERANCE OF LOCAL ERROR = 1E-7

MAXIMUM NUMBER OF

CORRECTION CYCLES ?3
```

SOLUTION
****


```
 X = .5   Y = .1484375
 X = 1   Y = .717346191
 X = 1.5   Y = 1.97937536
```

NO CONVERGENCE : STEP HALVED

H = .25

RE- START AT X = 0


```
 X = .25   Y = .0340169271
 X = .5   Y = .148699469
 X = .75   Y = .366958026
```

NO CONVERGENCE : STEP HALVED

H = .125

RE- START AT X = 0


```
 X = .125   Y = 8.14819336E-03
 X = .25   Y = .0340248281
 X = .375   Y = .0799904142
 X = .5   Y = .148721318
 X = .625   Y = .24324738
 X = .75   Y = .367003173
 X = .875   Y = .52388062
 X = 1   Y = .718289847
```

Program Listing

```
0   HOME
3   REM ==PREDICTOR/CORRECTOR==
10   DEF  FN F(X) = X + Y
15   REM   INPUT DATA
20   GOSUB 540
25   REM ==INITIALIZE VARIABLES AND STARTING VALUES==
30   GOSUB 1090
35   REM ==PREDICTOR==
40   GOSUB 1400
50 N = 0
55   REM ==CORRECTOR==
60   GOSUB 1430
65   REM ==CHECK FOR TOLERANCE==
70 D = Z(2) - Z(1)
80 D =  ABS (D) / H4
90   IF D < E2 THEN 140
95   REM ==DIFFERENCE TOO LARGE, APPLY CORRECTOR AGAIN==
96   REM  APPLY CORRECTOR AGAIN
100 N = N + 1
```

```
110   IF N >  = NO THEN 280
120  Z(1) = Z(2)
130   GOTO 60
135   REM ==CONVERGENCE ATTAINED==
140  XO = XO + H
150  YO = Z(2)
160   PRINT " X = ";XO;"   Y = ";YO
170  ZO = XO
180  WO = YO
190   IF  XO >  = B THEN 530
200   FOR I = 1 TO 3
210  F(I) = F(I + 1)
220   NEXT I
230  X = XO
240  Y = YO
250  F(4) =   FN F(X)
260   IF C$ = "Y" THEN 400
270   GOTO 40
280   IF C$ = "N" THEN 470
290   PRINT
300   PRINT "NO CONVERGENCE : STEP HALVED"
310   PRINT
320  H = H / 2
330   PRINT "H = ";H
340   PRINT
350   PRINT "RE- START AT X = ";ZO
360   PRINT
370  XO = ZO
380  YO = WO
390   GOTO 30
400   IF D > E1 THEN 40
410   PRINT
420   PRINT "TOO MUCH PRECISION : STEP DOUBLED"
430   PRINT
440  H = 2 * H
450   PRINT "H = ";H
460   GOTO 30
470   PRINT
480   PRINT "*** ERROR ***"
490   PRINT
500   PRINT "NO CONVERGENCE ATTAINED"
510   PRINT
520   PRINT "WITHIN SPECIFIED TOLERANCE"
530   END
535   REM ==PRINT TITLE PAGE==
540   PRINT
550   PRINT "PREDICTOR-CORRECTOR METHOD"
560   PRINT "FOR A FIRST ORDER EQUATION"
700   PRINT
710   PRINT
760   PRINT "ENTER INTERVAL OF INTEGRATION"
770   PRINT
775   REM ==INPUT DATA==
780   INPUT "A = ";A
790   INPUT "B = ";B
```

```
800    PRINT
810    PRINT "INITIAL CONDITION"
820    PRINT
830    INPUT "YA = ";YO
840    PRINT
850    PRINT "WANT STEPSIZE CONTROL (Y/N) ";
860    INPUT C$
870    PRINT
880    PRINT "INITIAL STEPSIZE"
890    PRINT
900    INPUT "H = ";HO
910    PRINT
920    INPUT "TOLERANCE OF LOCAL ERROR = ";E2
930    PRINT
940    PRINT "MAXIMUM NUMBER OF "
950    PRINT
960    PRINT "CORRECTION CYCLES ";
970    INPUT NO
980    PRINT
990    PRINT
1000    PRINT "SOLUTION"
1010    PRINT "********"
1020    PRINT
1025    REM ==INITIALIZE VARIABLES==
1030   H = HO
1040   XO = A
1050   ZO = XO
1060   WO = YO
1065    REM =='E1' IS LOWER BOUND FOR LOCAL ERROR==
1070   E1 = E2 / 1000
1080    RETURN
1085    REM ==INITIALIZE VALUES FOR STARTING PROCEDURE==
1090    PRINT
1100   X = XO
1110   Y = YO
1120   ZO = XO
1130   WO = YO
1140   H2 = H / 24
1150   H4 = H * 14
1160   F(1) =  FN F(X)
1165    REM ==4TH ORDER RUNGE-KUTTA METHOD TO COMPUTE STARTING VALUES==
1170   U = XO
1180   W = YO
1190    FOR I = 1 TO 3
1200   X = U
1210   Y = W
1220   K1 = H * ( FN F(X))
1230   X = U + 0.5 * H
1240   Y = W + 0.5 * K1
1250   K2 = H * ( FN F(X))
1260   Y = W + 0.5 * K2
1270   K3 = H * ( FN F(X))
1280   X = U + H
1290   Y = W + K3
1300   K4 = H * ( FN F(X))
```

```
1310 U = U + H
1320 W = W + (K1 + 2 * K2 + 2 * K3 + K4) / 6
1330  PRINT " X = ";U;"  Y = ";W
1340 Y = W
1350 F(I + 1) =  FN F(X)
1360  NEXT I
1370 XO = U
1380 YO = W
1390  RETURN
1396  REM ==4TH ORDER ADAMS - BASHFORT==
1397  REM ==METHOD (EXPLICIT)==
1400 Q = 55 * F(4) - 59 * F(3) + 37 * F(2) - 9 * F(1)
1410 Z(1) = YO + H2 * Q
1420  RETURN
1425  REM ==CORRECTED VALUE ==
1426  REM ==4TH ORDER ADAMS - MOULTON==
1430 X = XO + H
1440 Y = Z(1)
1450 F(5) =  FN F(X)
1460 Q = 9 * F(5) + 19 * F(4) - 5 * F(3) + F(2)
1470 Z(2) = YO + H2 * Q
1480  RETURN
```

References

Collatz, L. *The Numerical Treatment of Ordinary Differential Equations.* Springer Berlin, 1966.

Conte, Samuel D., and De Boor, Carl. *Elementary Numerical Analysis.* McGraw-Hill, 1972.

Gear, C. W. *Numerical Initial Value Problems in Ordinary Differential Equations.* Prentice-Hall, 1971.

Runge-Kutta Method for a System of Equations

This program will compute the solution of the Cauchy problem for the first order differential equation using a 4th order Runge-Kutta method:

$$\frac{dy}{dx} = f(Y(x), x) \quad X > A$$

$$Y(A) = Ya$$

Program Notes

To run this program, you must define the function $f(Y, X)$ in line 10.

Example

Compute the solution to the Cauchy problem:

$$\frac{dy}{dx} = x + y \quad x > 0$$

$$Y(0) = 0$$

```
10   DEF   FN F(X) = X + Y


]RUN
RUNGE-KUTTA METHOD FOR A

  SYSTEM OF EQUATIONS



>ENTER LIMITS OF INTEGRATION

 A = 0

 B = 1

 YA = 0

>INITIAL STEPSIZE

 H = .5

> LOCAL ERROR TOLERANCE

 E = 1E-7

>STEPSIZE CONTROL (Y/N) ?N
```

>SOLUTION
****

X = 0   Y = 0

 ** ERROR **   CONVERGENCY NOT
    ATTAINED WITHIN TOLERANCE
 FOR STEP SIZE H=.5 AT X=0

]RUN
RUNGE-KUTTA METHOD FOR A

   SYSTEM OF EQUATIONS

>ENTER LIMITS OF INTEGRATION

 A = 0

 B = 1

 YA = 0

>INITIAL STEPSIZE

 H = .5

.> LOCAL ERROR TOLERANCE

 E = 1E-7

>STEPSIZE CONTROL (Y/N) ?Y

>SOLUTION
****

X = 0   Y = 0

X = .0625   Y = 1.9944589E-03

X = .125   Y = 8.14845304E-03

X = .1875   Y = .0187302494

X = .25   Y = .0340254166

X = .3125   Y = .0543379411

X = .375   Y = .0799914145

```
X = .4375    Y = .111330299

X = .5    Y = .148721271

X = .5625    Y = .192554657

X = .625    Y = .243245957

X = .6875    Y = .301237469

X = .75    Y = .367000016

X = .8125    Y = .441034787

X = .875    Y = .523875294

X = .9375    Y = .616089457

X = 1    Y = .718281828

>END OF EXECUTION
```

Program Listing

```
1    HOME
10   DEF  FN F(X) = X + Y
20   PRINT "RUNGE-KUTTA METHOD FOR A "
25   PRINT
30   PRINT "  SYSTEM OF EQUATIONS   "
110   PRINT
130   PRINT
135   REM ==INPUT DATA==
140   PRINT
150   PRINT ">ENTER LIMITS OF INTEGRATION"
160   PRINT
170   INPUT " A = ";A
180   PRINT
190   INPUT " B = ";B
200   PRINT
210   INPUT " YA = ";Y0
220   PRINT
230   PRINT ">INITIAL STEPSIZE "
240   PRINT
250   INPUT " H = ";H0
260   PRINT
270   PRINT "> LOCAL ERROR TOLERANCE"
280   PRINT
290   INPUT " E = ";E1
300   PRINT
310   PRINT ">STEPSIZE CONTROL (Y/N) ";
320   INPUT C$
330   PRINT
340   PRINT ">SOLUTION"
350   PRINT "********"
360   PRINT
```

```
370     PRINT
377     REM ==INITIALIZE VARIABLES==
380  X0 = A
390  H = H0
400  E5 = E1 / 100
405     REM ==OUTPUT RESULTS==
410     PRINT
420     PRINT "X = ";X0; SPC( 3);"Y = ";Y0
430     GOSUB 480
435     REM ==CHECK FOR END OF INTERVAL==
440     IF X0 < = B THEN 410
450     PRINT
460     PRINT ">END OF EXECUTION"
470     GOTO 1000
475     REM ==INTERGRATION ROUTINE==
477     REM ==NEXT POINT W/CURRENT STEP SIZE==
480  U = X0
490  W = Y0
500  T = H
510     GOSUB 800
520  Q1 = W
530  T2 = H
535     REM ==NEXT POINT W/ HALF STEP SIZE==
540  U = X0
550  W = Y0
560  T = T2 / 2
570     FOR I = 1 TO 2
580     GOSUB 800
590  Z(I) = W
600     NEXT I
610  Q2 = Z(2)
615     REM ==CHECK FOR CONVERGENCE==
620  Q3 = (Q2 - Q1) / 15
630  D =  ABS (Q3) / T2
640     IF D < E1 THEN 740
645     REM ==OUT OF TOLERANCE==
650     IF C$ = "Y" THEN 710
655     REM ==ABORT EXECUTION==
660     PRINT
670     PRINT " ** ERROR **   CONVERGENCY NOT"
680     PRINT "    ATTAINED WITHIN TOLERANCE"
690     PRINT " FOR STEP SIZE H=";H;" AT X=";X0
695     GOTO 1000
700     PRINT
704     REM ==CONVERGENCE NOT ATTAINED-HALF STEP SIZE==
710  Q1 = Z(1)
720  T2 = T2 / 2
725     REM ==ITERATE AGAIN==
730     GOTO 540
735     REM ==CONVERGENCE ATTAINED==
736     REM ==RICHARDSON EXTRAPOLATION==
740  Y0 = Q2 + Q3
750  X0 = X0 + T2
760     IF C$ = "N" THEN 800
765     REM ==DOUBLE STEP SIZE IF TOO SMALL==
```

```
770   IF D > E5 THEN 800
780 H = 2 * H
790   RETURN
800 X = U
805   REM ==4TH ORDER RUNGE-KUTTA=
810 Y = W
820 K1 =   FN F(X)
830 K1 = T * K1
840 X = U + T / 2
850 Y = W + K1 / 2
860 K2 =   FN F(X)
870 K2 = T * K2
880 Y = W + K2 / 2
890 K3 =   FN F(X)
900 K3 = T * K3
910 X = U + T
920 Y = W + K3
930 K4 =   FN F(X)
940 K4 = T * K4
950 V1 = K1 + 2 * K2 + 2 * K3 + K4
960 W = W + V1 / 6
970 U = X
980   RETURN
1000   END
```

References

Gear, C. W. *Numerical Initial Value Problems in Ordinary Differential Equations.* Prentice-Hall, 1971.

Conte, Samuel D., and De Boor, Carl. *Elementary Numerical Analysis.* McGraw-Hill, 1972.

8
Numerical Integration

Integral Evaluation by Modified Simpson's Method

This program provides a means of limiting cumulative machine errors based on Simpson's formula for evaluation of a definite integral.

$$\int_A^B f(X)DX = \frac{H}{3}\left[f(X_1) = 4f(X_2) + 2f(X_3) + \ldots 4f(X_n0 + f(X_{n+1})\right]$$

$$-\frac{H^4(B-A)f^{iv}(X\times)}{180}$$

where: $H = (B-A)/n$
n is an even integer
$A < X\times < B$

Integral = Principal part + Error

$A = X_1 \quad X_2 \quad X_3 \qquad X_n \quad X_{n+1} = B$

Example

Let $f(X) = 1/X; A = 1; B = 2; E = 10^{-5}$. The exact $f^{iv} = 24 X^{-5}$ and with $M = 1.5, f^{iv}(1.5) = 3.16049$, determine the most accurate answer.

```
RUN
INTEGRAL EVALUATION BY MEANS OF

     A MODIFIED SIMPSON


KEY IN LOWER LIMIT : UPPER LIMIT 2,5
KEY IN THE TOLERANCE 1E-6



FOR A TOLERANCE OF 1E-06 , I(12) = 2.22540032



DO YOU WANT A NEW TOLERANCE
(1 = YES, 0 = NO) ?0
```

Program Listing

```
100  HOME
110  PRINT "INTEGRAL EVALUATION BY MEANS OF "
113  PRINT
```

```
115   PRINT "        A MODIFIED SIMPSON"
120   PRINT
130   PRINT
180   INPUT "KEY IN LOWER LIMIT : UPPER LIMIT ";A,B
190   INPUT "KEY IN THE TOLERANCE ";E
200 M = (A + B) / 2
205 H1 = (B - A) / 40
210   GOSUB 390
218   REM ==EVALUATE PRINCIPAL TERMS OF DEFINITE INTEGRAL==
220 N = ((B - A) ^ 5 * D / E / 180) ^ .25
230 N = 2 *  INT (N / 2 + 1)
240 S = 0
243 X = A
247 H = (B - A) / N
250   FOR I = 1 TO N / 2
260   GOSUB 370
265 S = S + F
270 X = X + H
273   GOSUB 370
277 S = S + 4 * F
280 X = X + H
283   GOSUB 370
287 S = S + F
290   NEXT I
300   PRINT : PRINT : PRINT
310   PRINT "FOR A TOLERANCE OF ";E;" , ";"I(";N;") = ";S * H / 3
320   PRINT : PRINT
330   PRINT "DO YOU WANT A NEW TOLERANCE"
335   PRINT "(1 = YES, 0 = NO) ";
340   INPUT W
350   IF W = 1 THEN 190
360   GOTO 470
368   REM ==SUBROUTINE WHICH DEFINES USER FUNCTION==
370 F = X /  SQR (9 + X * X)
380   RETURN
388   REM ==SUBROUTINE TO EVALUATE ERROR==
390 D = 0
400 X = M - 2 * H1
403   GOSUB 370
407 D = D + F
410 X = M - H1
413   GOSUB 370
417 D = D - 4 * F
420 X = M
423   GOSUB 370
427 D = D + 6 * F
430 X = M + H1
433   GOSUB 370
437 D = D - 4 * F
440 X = M + 2 * H1
443   GOSUB 370
447 D = D + F
450 D =  ABS (D) / H1 ^ 4
460   RETURN
470   END
```

References

Jennings, W. *First Course in Numerical Analysis.*

Gaussian Evaluation of a Definite Integral

This program, like the Modified Simpson method, is used to limit the amount of machine-introduced error in the evaluation of integrals. In the Gauss method, however, the interval (A,B) is transformed to an interval ($-1,1$) and the function is evaluated at symmetrically arranged argument values which are then multiplied by the corresponding weight factors.

Program Notes

It will be necessary to enter your actual function at line 150 of the program.

Example

Find the correct value for the following:

$$F = 1/X$$
$$(1,2)$$

```
RUN

GAUSSIAN EVALUATION

OF A DEFINITE INTEGRAL


KEY IN LOWER LIMIT A,UPPER LIMIT B 1,2


THE VALUE OF THE INTEGRAL IS .693147129


DO YOU WANT NEW LIMITS
(YES = 1, NO = 0)?1
KEY IN LOWER LIMIT A,UPPER LIMIT B 3,6


THE VALUE OF THE INTEGRAL IS .693147129


DO YOU WANT NEW LIMITS
(YES = 1, NO = 0)?0
```

Program Listing

```
100    HOME
110    PRINT
120    PRINT "GAUSSIAN EVALUATION"
123    PRINT
125    PRINT "OF A DEFINITE INTEGRAL"
130    PRINT
140    PRINT
150    DEF  FN G(X) = 1 / X
160    PRINT
170    INPUT "KEY IN LOWER LIMIT A,UPPER LIMIT B ";A,B
180 C = (B - A) / 2
185 D = (B + A) / 2
190 S = 0
200    FOR I = 1 TO 6
210    READ X,W
220 S = S + W * ( FN G(C * ( - X) + D) +  FN G(C * X + D))
230    NEXT I
240 V = C * S
250    PRINT
260    PRINT
270    PRINT "THE VALUE OF THE INTEGRAL IS ";V: PRINT : PRINT : PRINT
271    PRINT "DO YOU WANT NEW LIMITS"
272    PRINT "(YES = 1, NO = 0)";
273    INPUT QQ
274    IF QQ = 1 THEN  RESTORE : GOTO 170
280    GOTO 350
290    DATA  .981561,.0471753
300    DATA  .904117,.106939
310    DATA  .769903,.160078
320    DATA  .587318,.203167
330    DATA  .367831,.233493
340    DATA  .125333,.249147
350    END
```

References

Froberg. *Introduction to Numerical Analysis.*

Fredholm Integral Equation

This program computes the solution of the Fredholm integral equation of the second kind:

$$f(X) + \lambda \int_A^B K(X,T) \times f(T) \times DT = G(X)$$

in which A, B, λ, K, and G are known. (In some cases the λ can be an eigenvalue — usually incorporated into the K.)

The method of solution used here is the Simpson closed form, in which the error will be $0(H^4)$ and the number of subdivisions must be even.

$$\text{Let } f_i = f(X_i) \,;\, G_i = G(X_i) \,;\, K_{ij} = K(X_i,T_j) \,;\, H = (B - A)/N \,;\, N = 2,4,6,\ldots$$

Example

Find the exact solution for:

$$f(X) + \int_1^3 X^2 T \times f(T) \times DT = 21X^2$$

```
RUN
FREDHOLM INTEGRAL EQUATION


   KEY IN VALUES OF A,B,N 2,7,10



              RUNNING


       X-VALUE      F(X)-VALUE

       2           -5.0899716
       2.5           -3.65299616
       3           -2.3421412
       3.5           -1.19887134
       4           -.235127169
       4.5          .555090189
       5          1.18824349
```

5.5	1.68661869
6	2.07539644
6.5	2.38069077
7	2.62812001

Program Listing

```
10    HOME
20    DIM A(40,40),C(40),X(40)
30    PRINT "FREDHOLM INTEGRAL EQUATION"
40    PRINT
50    PRINT
90    PRINT
100   INPUT "  KEY IN VALUES OF A,B,N ";A,B,N
110   HOME : PRINT : PRINT : PRINT : PRINT
120   PRINT "                    RUNNING"
125   PRINT : PRINT
130   H = (B - A) / N
140   X = A
145   T = A
150   FOR I = 1 TO N + 1
160   GOSUB 900
170   A(I,1) = K
180   IF I < > 1 THEN 200
190   A(I,1) = 1 + K
200   X = X + H
210   NEXT I
220   T = A + H
230   FOR J = 2 TO N
240   X = A
250   FOR I = 1 TO N + 1
260   GOSUB 900
270   IF INT (J / 2) * 2 < > J THEN 330
280   A(I,J) = 4 * K
290   IF I = J THEN 310
300   GOTO 370
310   A(I,J) = 1 + 4 * K
320   GOTO 370
330   A(I,J) = 2 * K
340   IF I = J THEN 360
350   GOTO 370
360   A(I,J) = 1 + 2 * K
370   X = X + H
380   NEXT I
390   T = T + H
400   NEXT J
410   X = A
415   T = B
420   FOR I = 1 TO N + 1
430   GOSUB 900
440   A(I,N + 1) = K
450   GOSUB 920
460   C(I) = G
470   IF I < > N + 1 THEN 490
```

```
480 A(I,N + 1) = 1 + K
490 X = X + H
500  NEXT I
510  FOR Z = 1 TO N + 1
520 A(Z,N + 2) = C(Z)
530  NEXT Z
540 N = N + 1
543 M = N + 1
547 L = N - 1
550  FOR K = 1 TO L
560 JJ = K
563 BI =  ABS (A(K,K))
567 KP = K + 1
570  FOR I = KP TO N
580 A1 =  ABS (A(I,K))
590  IF (BI - A1) >  = 0 THEN 610
600 BI = A1
605 JJ = I
610  NEXT I
620  IF (JJ - K) = 0 THEN 670
630  FOR J = K TO M
640 TE = A(JJ,J)
650 A(JJ,J) = A(K,J)
660 A(K,J) = TE
665  NEXT J
670  FOR I = KP TO N
680 QU = A(I,K) / A(K,K)
690  FOR J = KP TO M
700 A(I,J) = A(I,J) - QU * A(K,J)
710  NEXT J
715  NEXT I
720  FOR I = KP TO N
730 A(I,J) = 0
733  NEXT I
737  NEXT K
740 X(N) = A(N,M) / A(N,N)
750  FOR NN = 1 TO L
760 SU = 0
763 I = N - NN
767 IP = I + 1
770  FOR J = IP TO N
780 SU = SU + A(I,J) * X(J)
785  NEXT J
790 X(I) = (A(I,M) - SU) / A(I,I)
795  NEXT NN
800  HOME
810  PRINT "             X-VALUE     F(X)-VALUE"
820  PRINT
840 X = A
850  FOR I = 1 TO N
860  PRINT "              ";X;"          ";X(I)
870 X = X + H
880  NEXT I
890  GOTO 940
900 K = 2 * H *  LOG (X * X + T * T) / 3
```

```
910   RETURN
920 G = 4 +   SIN (.5 * X)
930   RETURN
940   END
```

References

Froberg. *Introduction to Numerical Analysis.*
Scheid. *Numerical Analysis.*

9
Fourier Analysis

Fourier Series Analysis of
Periodic Functions

Fourier Series Expansion of Piecewise
Linear Periodic Functions

Fourier Series Analysis of Periodic Functions

This is a general program for the evaluation of the coefficients of the Fourier series expansions of periodic functions:

$$f(X) = \frac{A_o}{2} + A_1 \cos. W_1 X + A_2 \cos. W_2 X + \ldots + A_n \cos W_n X + \ldots$$

$$+ B_1 \sin. W_1, X + B_2 \sin W_2 X + \ldots + B_n \cos. W_n X + \ldots$$

where $W_n = \frac{2\pi N}{T}$, T being the period of the function

Program Notes

$f(X)$ can be defined using a DEF FN F(X) statement on line 10 or piecewise (in which case the coordinates of the vertices must be input).

The maximum number of Fourier coefficients that can be computed using this program is limited to 10. To modify this, change line 910 as follows:

$$910\ \text{DIM}\ Q(N),\ R(N)$$

where N is the maximum number of coefficients.

The number of intervals for the numerical integration is set in line 90:

$$90\ \text{M=N8*J}$$

in which J is the order of the coefficients + 1, and $N8$ is set equal to 20 (in line 1150).

In order to reduce the computation time (when computing more than five coefficients) either of these statements can be modified as needed.

Example

Compute the first five coefficients of the Fourier expansion of the sawtooth function shown below:

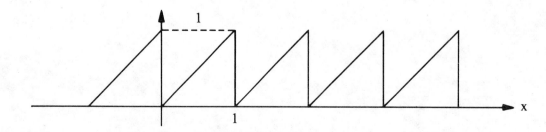

RUN

FOURIER SERIES ANALYSIS OF

PERIODIC FUNCTIONS

DEFINITION OF FUNCTION F:

    TYPE 1 : DEFINED IN LINE # 10

    TYPE 2 : PIECEWISE LINEAR

    TYPE = 2

INTERVAL OF EXPANSION

    X0 < X < X1

    X0 = 0

    X1 = 1

TYPE OF EXPANSION

    TYPE 0 : GENERAL

    TYPE 1 : COSINE ONLY

    TYPE 2 : SINE ONLY

    TYPE = 0

DESIRED NUMBER OF TERMS

IN EXPANSION = 5

F(X) DEFINED AS PIECEWISE LINEAR

NUMBER OF SEGMENTS = 1

ENTER END POINT COORDINATES X,Y

(IN ORDER OF INCREASING X)

POINT   1

X = 0
Y = 0

POINT   2

X = 1
Y = 1

```
PERIOD = 1

FOURIER COEFFICIENTS
********************


N = 0

W = 0     W/TWOPI = 0

A = 1     B = 0


N = 1

W = 6.28318531     W/TWOPI = 1

A = 0     B = -.318310966


N = 2

W = 12.5663706     W/TWOPI = 2

A = 0     B = -.159156653


N = 3

W = 18.8495559     W/TWOPI = 3

A = 0     B = -.106105125


N = 4

W = 25.1327412     W/TWOPI = 4

A = 0     B = -.0795792491


N = 5

W = 31.4159266     W/TWOPI = 5

A = 0     B = -.0636636531
```

Program Listing

```
1    HOME
10   REM ==FOURIER SERIES ANALYSIS OF PERIODIC FUNCTIONS==
15   REM ==INPUT DATA==
20   GOSUB 330
25   REM ==INITIALIZE VARIABLES==
30   GOSUB 1090
34   REM ==COMPUTE L + 1 COEFFICIENTS==
40 L1 = L + 1
50   FOR J = 1 TO L1
60 J1 = J - 1
70 W = W0 * J1
80 G = F0 * J1
85   REM ==M IS NUMBER OF INTERVALS FOR NUMERICAL INTEGRATION==
90 M = N8 * J
100 M1 = M - 1
110 H = (X1 - X0) / M
120 H2 = 2 * H
125   REM ==COSINE COEFFICIENT A==
130   IF K8 = 2 THEN 190
140   ON K9 GOSUB 1760,1360
150 A = U * S / T
160   IF  ABS (A) > 1E - 8 THEN 180
170 A = 0
175   REM ==SINE COEFFICIENT B==
180   IF K8 = 1 THEN 230
190   ON K9 GOSUB 1950,1560
200 B = U * S / T
210   IF  ABS (B) > 1E - 8 THEN 230
220 B = 0
225   REM ==PRINT RESULTS==
230   PRINT
240   PRINT "N = ";J1
250   PRINT
260   PRINT "W = ";W;"     W/TWOPI = ";G
270   PRINT
280   PRINT "A = ";A;"     B = ";B
290   PRINT
300   PRINT
310   NEXT J
320   GOTO 2140
325   REM ==INPUT ROUTINE==
330   PRINT
340   PRINT "FOURIER SERIES ANALYSIS OF"
350   PRINT
360   PRINT "PERIODIC FUNCTIONS": PRINT : PRINT
580   PRINT "DEFINITION OF FUNCTION F:"
590   PRINT
600   PRINT "    TYPE 1 : DEFINED IN LINE # 10"
610   PRINT
620   PRINT "    TYPE 2 : PIECEWISE LINEAR"
630   PRINT
640   INPUT "    TYPE = ";K9
650   PRINT
660   PRINT "INTERVAL OF EXPANSION"
```

```
670    PRINT
680    PRINT "    X0 < X < X1 "
690    PRINT
700    INPUT "    X0 = ";X0
710    PRINT
720    INPUT "    X1 = ";X1
730    PRINT
740    PRINT
750    PRINT "TYPE OF EXPANSION"
760    PRINT
770    PRINT "    TYPE 0 : GENERAL "
780    PRINT
790    PRINT "    TYPE 1 : COSINE ONLY"
800    PRINT
810    PRINT "    TYPE 2 : SINE ONLY"
820    PRINT
830    INPUT "    TYPE = ";K8
840    PRINT
850    PRINT "DESIRED NUMBER OF TERMS"
860    PRINT
870    INPUT "IN EXPANSION = ";L
880    IF K9 = 1 THEN 1080
890    PRINT
900    PRINT
910    DIM Q(10),R(10)
920    PRINT "F(X) DEFINED AS PIECEWISE LINEAR"
930    PRINT
940    INPUT "NUMBER OF SEGMENTS = ";N1
950    PRINT
960    PRINT "ENTER END POINT COORDINATES X,Y"
970    PRINT
980    PRINT "(IN ORDER OF INCREASING X)
990    PRINT
1000   N2 = N1 + 1
1010    FOR I = 1 TO N2
1020    PRINT "POINT   ";I
1030    PRINT
1040    INPUT "X = ";Q(I)
1050    INPUT "Y = ";R(I)
1060    PRINT
1070    NEXT I
1080    RETURN
1085    REM ==INITIALIZATION==
1090   P1 = 3.1415926538
1100   P2 = 2 * P1
1110   U = 2
1120   T = X1 - X0
1130   A = 0
1140   B = 0
1150   N8 = 20
1160    IF K8 = 0 THEN 1200
1165    REM ==DOUBLE PERIOD FOR HALF EXPANSIONS==
1170   T = 2 * T
1180   U = 2 * U
1190    PRINT
```

```
1200   PRINT
1210   PRINT "PERIOD = ";T
1220   PRINT
1230   PRINT
1240   PRINT "FOURIER COEFFICIENTS"
1250   PRINT "********************"
1260   PRINT
1270   PRINT
1280 F0 = 1 / T
1290 W0 = P2 * F0
1300   IF K9 = 1 THEN 1330
1305   REM ==F(X) PIECEWISE LINEAR==
1310   DEF  FN F(X) = Y0 + P * (X - X0)
1320   GOTO 1350
1330 Y0 =  FN F(X0)
1340 Y1 =  FN F(X1)
1350   RETURN
1355   REM ==COSINE COEF. FOR F(X) PIECEWISE LINEAR==
1360 A1 = 0
1370 J3 = 0
1380 J3 = J3 + 1
1390   IF J3 > N1 THEN 1540
1400 J4 = J3 + 1
1410 X0 = Q(J3)
1420 X1 = Q(J4)
1430 E = X1 - X0
1440   IF E = 0 THEN 1380
1450 Y0 = R(J3)
1460 Y1 = R(J4)
1470 D = Y1 - Y0
1480 P = D / E
1490 H = E / M
1500 H2 = 2 * H
1505   REM ==PERFORM INTEGRATION==
1510   GOSUB 1760
1520 A1 = A1 + S
1530   GOTO 1380
1540 S = A1
1550   RETURN
1555   REM  ==SINE COEF. FOR F(X) PIECEWISE LINEAR==
1560 B1 = 0
1570 J3 = 0
1580 J3 = J3 + 1
1590   IF J3 > N1 THEN 1740
1600 J4 = J3 + 1
1610 X0 = Q(J3)
1620 X1 = Q(J4)
1630 E = X1 - X0
1640   IF E = 0 THEN 1580
1650 Y0 = R(J3)
1660 Y1 = R(J4)
1670 D = Y1 - Y0
1680 P = D / E
1690 H = E / M
1700 H2 = 2 * H
1705   REM ==PERFORM INTEGRATION==
```

```
1710   GOSUB 1950
1720 B1 = B1 + S
1730   GOTO 1580
1740 S = B1
1750   RETURN
1755   REM ==NUMERICAL INTEGRATION (SIMPSON'S RULE) FOR COSINE
COEFFICIENT A==
1760 S1 = 0
1770 S2 = 0
1780 T1 = X0
1790 T2 = T1 + H
1800   FOR I = 1 TO M1 STEP 2
1810 D1 =   COS (W * T1)
1820 D2 =   COS (W * T2)
1830 F1 =   FN F(T1)
1840 F2 =   FN F(T2)
1850 S1 = S1 + D1 * F1
1860 S2 = S2 + D2 * F2
1870 T1 = T1 + H2
1880 T2 = T2 + H2
1890   NEXT I
1900 Z0 = Y0 *   COS (W * X0)
1910 Z1 = Y1 *   COS (W * X1)
1920 S = 2 * S1 + 4 * S2 - Z0 + Z1
1930 S = S * H / 3
1940   RETURN
1945   REM ==NUMERICAL INTEGRATION (SIMPSON'S RULE) FOR SINE
COEFFICIENT B==
1950 S1 = 0
1960 S2 = 0
1970 T1 = X0
1980 T2 = T1 + H
1990   FOR I = 1 TO M1 STEP 2
2000 D1 =   SIN (W * T1)
2010 D2 =   SIN (W * T2)
2020 F1 =   FN F(T1)
2030 F2 =   FN F(T2)
2040 S1 = S1 + D1 * F1
2050 S2 = S2 + D2 * F2
2060 T1 = T1 + H2
2070 T2 = T2 + H2
2080   NEXT I
2090 Z0 = Y0 *   SIN (W * X0)
2100 Z1 = Y1 *   SIN (W * X1)
2110 S = 2 * S1 + 4 * S2 - Z0 + Z1
2120 S = S * H / 3
2130   RETURN
2140   END
```

Reference

Tolstov, G.P. *Fourier Series.* Prentice-Hall, 1962.

Fourier Series Expansion of Piecewise Linear Periodic Functions

This program computes Fourier series expansions on those functions $f(X)$ which are piecewise-linear. In this instance, the Fourier coefficients can be computed in closed form, segment by segment. Since the need for numerical integration is therefore eliminated, this program will run much more quickly than the general form for Fourier series expansions.

Program Notes

The maximum number of coefficients that can be computed is currently set to 10. To modify the program, alter the DIM statement in line 10 as follows:

```
10 DIM A(N), B(N)
```

where N is the new number of coefficients.

Example

Compute the coefficients of the Fourier series expansion of:

$$f(X) = E^{-x}$$

over the range:

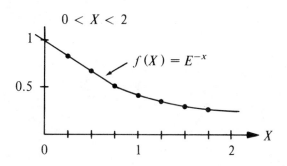

RUN

```
FOURIER SERIES EXPANSION OF

PIECE-WISE LINEAR PERIODIC

        FUNCTION


PERIOD = 1

NUMBER OF SEGMENTS = 1
```

```
DESIRED NUMBER OF TERMS

IN EXPANSION = 5


ENTER VERTEX COORDINATES

     X , F(X)

(IN ORDER OF INCREASING X)

VERTEX 1?0,0

VERTEX 2?1,1

FOURIER COEFFICIENTS
********************

>N = 0   W = 0
A = 1

>N = 1   W = 6.28318531   W/TWOPI = 1
A = 0     B = -.318309886

>N = 2   W = 12.5663706   W/TWOPI = 2
A = 0     B = -.159154943

>N = 3   W = 18.8495559   W/TWOPI = 3
A = 0     B = -.106103295

>N = 4   W = 25.1327412   W/TWOPI = 4
A = 0     B = -.0795774712

>N = 5   W = 31.4159266   W/TWOPI = 5
A = 0     B = -.063661977
```

Program Listing

```
0    HOME
5    PRINT "FOURIER SERIES EXPANSION OF "
6    PRINT "PIECEWISE LINEAR PERIODIC FUNCTIONS"
8    REM ==INPUT DATA==
10   DIM A(10),B(10)
20   GOSUB 60
26   REM ==COMPUTE COEFFICIENTS==
30   GOSUB 450
40   GOSUB 900
50   GOTO 1180
60   PRINT
100  PRINT
110  PRINT "        FUNCTION"
120  PRINT
280  PRINT
```

```
290   INPUT "PERIOD = ";T
300   PRINT
310   INPUT "NUMBER OF SEGMENTS = ";N
320   PRINT
330   PRINT "DESIRED NUMBER OF TERMS"
340   PRINT
350   INPUT "IN EXPANSION = ";M
355   REM ==INITIALIZE VARIABLES==
360 P1 = 3.141592654
370 P2 = 2 * P1
380 W0 = P2 / T
390 A0 = 0
400   FOR I = 1 TO M
410 A(I) = 0
420 B(I) = 0
430   NEXT I
440   RETURN
445   REM ==INPUT SEGMENT DATA==
450   PRINT
460   PRINT
470   PRINT "ENTER VERTEX COORDINATES"
480   PRINT
490   PRINT "     X , F(X) "
500   PRINT
510   PRINT "(IN ORDER OF INCREASING X)"
520   PRINT
530   PRINT "VERTEX 1";
540   INPUT X0,F0
550 K = 0
560 K = K + 1
565   REM ==CHECK FOR LAST SEGMENT==
570   IF K > N THEN 890
580 K1 = K + 1
590   PRINT
600   PRINT "VERTEX ";K1;
610   INPUT X1,F1
620   IF X1 = X0 THEN 860
625   REM ==COMPUTE COEFFICIENTS==
630 A0 = A0 + (X1 - X0) * (F1 + F0)
640 P = (F1 - F0) / (X1 - X0)
650 Q = F0 - P * X0
660 W = 0
670   FOR I = 1 TO M
680 W = W + W0
690 Z0 = W * X0
700 Z1 = W * X1
710 C0 =   COS (Z0)
720 C1 =   COS (Z1)
730 S0 =   SIN (Z0)
740 S1 =   SIN (Z1)
750 C2 = C1 - C0
760 C3 = X1 * C1 - X0 * C0
770 S2 = S1 - S0
780 S3 = X1 * S1 - X0 * S0
790 E0 = P / (W * W)
```

```
800 E1 = Q / W
810 A1 = E0 * (C2 + W * S3) + E1 * S2
820 B1 = E0 * (S2 - W * C3) - E1 * C2
830 A(I) = A(I) + A1
840 B(I) = B(I) + B1
850  NEXT I
860 X0 = X1
870 F0 = F1
880  GOTO 560
890  RETURN
895  REM ==OUTPUT RESULTS==
900  PRINT
910  PRINT "FOURIER COEFFICIENTS"
920  PRINT "********************"
930  PRINT
940 T2 = 2 / T
950 A0 = A0 / T
960 W = 0
970  PRINT ">N = 0   W = 0"
980  PRINT
990  PRINT "A = ";A0
1000  PRINT
1010  FOR I = 1 TO M
1020 W = W + W0
1030 F = W / P2
1040 A1 = A(I) * T2
1050 B1 = B(I) * T2
1060  IF  ABS (A1) > 1E - 8 THEN 1080
1070 A1 = 0
1080  IF  ABS (B1) > 1E - 8 THEN 1100
1090 B1 = 0
1100  PRINT ">N = ";I;"   W = ";W;
1110  PRINT "   W/TWOPI = ";F
1130  PRINT "A = ";A1; SPC( 3);"B = ";B1
1140  PRINT
1160  NEXT I
1170  RETURN
1180  END
```

References

Tolstov, G. P. *Fourier Series.* Prentice-Hall, 1962.

Churchill, R. V. *Fourier Series and Boundary Value Problems.* McGraw-Hill, 1963.

10
Structural Analysis

State of Stresses at a Point

This program will compute all the relevant quantities related to a second-rank symmetric tensor in a two- or three-dimensional vector space.

The solution is computed from the equation:

$$S = \sigma - PI$$

where:

σ is the given tensor

$P = (\sigma_{CC} + \sigma_{YY} + \sigma_{ZZ}) / 3$

I is the identity tensor

The inputs required for the program are:

$$\sigma_{XX}, \sigma_{YY}, \sigma_{ZZ}, \sigma_{XZ}, \sigma_{YZ}, \sigma_{XY}$$

with respect to the Cartesian system of reference.

The program will output:

· Invariants *J1, J2, J3*

· Principal values and principal directions for any direction defined by a vector N (the stress vector acting on the plane of normal vector N).

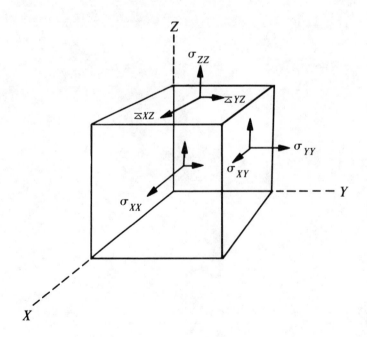

Example

Run for stresses using the following sigmas:

$$\sigma_{XX} = -10$$
$$\sigma_{YY} = -10$$
$$\sigma_{ZZ} = -10$$
$$\sigma_{XY} = 0$$
$$\sigma_{XZ} = 0$$
$$\sigma_{YZ} = 0$$

```
RUN
STATE OF STRESSES AT A POINT

>SYMMETRIC TENSOR COMPONENTS

   SIGMA XX = ?-10

   SIGMA YY = ?-10

   SIGMA ZZ = ?-10

   SIGMA XY = ?0

   SIGMA XZ = ?0

   SIGMA YZ = ?0
>GIVEN TENSOR INVARIANTS

   J1 = -10

   J2 = -300

   J3 = -1000


   PRESS ANY KEY TO CONTINUE


>DEVIATOR TENSOR INVARIANTS

   J1 = 0

   J2 = 0

   J3 = 0


   PRESS ANY KEY TO CONTINUE

>PRINCIPAL VALUES AND DIRECTIONS
```

```
SIGMA(1)=SIGMA(2)=SIGMA(3)= -10


**SINGULAR POINT**

ANY DIRECTION IS PRINCIPAL


WANT COMPONENTS FOR
ANY OTHER DIRECTION (Y/N)?Y

>INPUT DIRECTION

  N(1) = ?3

  N(2) = ?4

  N(3) = ?0

>COMPONENTS ON  N  DIRECTION

  SIG(1) = -6

  SIG(2) = -8

  SIG(3) = 0

WANT COMPONENTS FOR
ANY OTHER DIRECTION (Y/N)?Y

>INPUT DIRECTION

  N(1) = ?0

  N(2) = ?1

  N(3) = ?1

>COMPONENTS ON  N  DIRECTION

  SIG(1) = 0

  SIG(2) = -7.07106781

  SIG(3) = -7.07106781

WANT COMPONENTS FOR
ANY OTHER DIRECTION (Y/N)?N
END OF EXECUTION
```

Program Listing

```
10    HOME
20    PRINT "STATE OF STRESSES AT A POINT"
30    PRINT
40    PRINT
90    GOTO 3000
105   REM  ==MAIN PROGRAM==
110 E1 = 1E - 5
120 PI = 3.141592654
130   DIM S(6),D(3),P(3),A(3,3)
140   DATA  XX,YY,ZZ,XY,XZ,YZ
150   PRINT
160   PRINT ">SYMMETRIC TENSOR COMPONENTS"
170   FOR I = 1 TO 6
180   PRINT
190   READ S$
200   PRINT "  SIGMA ";S$;" = ";
210   INPUT S(I)
220   NEXT I
230   PRINT
240   PRINT
250   PRINT ">GIVEN TENSOR INVARIANTS"
255   REM =COMPUTE INVARIANTS==
260   GOSUB 500
265   REM ==OUTPUT==
270   GOSUB 1780
280   GOSUB 1870
290   PRINT ">DEVIATOR TENSOR INVARIANTS"
300 P = (S(1) + S(2) + S(3)) / 3
305   REM  ==COMPUTE DEVIATOR==
310   FOR I = 1 TO 3
320 S(I) = S(I) - P
330   NEXT I
340   GOSUB 500
350   GOSUB 1780
360   GOSUB 1870
370   PRINT ">PRINCIPAL VALUES AND DIRECTIONS"
380   PRINT
385   REM ==MAIN SUBROUTINE==
390   GOSUB 630
400   PRINT
410   PRINT
420   PRINT "WANT COMPONENTS FOR"
430   PRINT "ANY OTHER DIRECTION (Y/N)";
440   INPUT Z$
450   IF Z$ = "N" THEN 480
460   GOSUB 1930
470   GOTO 420
480   PRINT "END OF EXECUTION"
490   GOTO 3000
495   REM  ==INVARIANTS EVALUATION==
500 J1 = (S(1) + S(2) + S(3)) / 3
510 J2 = 0
520   FOR I = 1 TO 2
530 L = I + 1
```

```
540    FOR J = L TO 3
550 K = I + J + 1
560 J2 = J2 - S(I) * S(J) + S(K) * S(K)
570    NEXT J
580    NEXT I
590 J3 = (S(4) * S(6) - S(2) * S(5)) * S(5)
600 J3 = J3 + (S(5) * S(6) - S(3) * S(4)) * S(4)
610 J3 = J3 + (S(2) * S(3) - S(6) * S(6)) * S(1)
620    RETURN
625    REM   ==MAIN SUBROUTINE==
630 T =   SQR (J2)
640    IF   ABS (T) > E1 THEN 690
650    FOR J = 1 TO 3
660 D(J) = 0
670    NEXT J
680    GOTO 810
685    REM   ==W-ANGLE EVALUATION==
690 W =   - 3 *  SQR (3) * J3 / (2 * T * J2)
700    IF   ABS ( ABS (W) - 1) > E1 THEN 760
710    IF W > 0 THEN 740
720 W = PI / 3
730    GOTO 770
740 W = 0
750    GOTO 770
755    REM   ==COMPUTE ARCCOS(W)/3 IF W#0 AND W#PI/3==
760 W = (PI / 2 -  ATN (W /  SQR (1 - W * W))) / 3
770 CO = 2 * T /  SQR (3)
775    REM   ==PRINCIPAL VALUES OF DEVIATOR==
780 D(1) = CO *  COS (W - PI / 3)
790 D(2) = CO *  COS (W + PI / 3)
800 D(3) =   - CO *  COS (W)
805    REM   ==PRINCIPAL VALUES GIVEN TENSOR==
810    FOR I = 1 TO 3
820 P(I) = D(I) + P
830    NEXT I
835    REM =="SINGULARITY" OF THE PT==
840 R = 0
850    FOR J = 1 TO 2
860 I = J + 1
870    IF   ABS (D(I) - D(J)) > E1 THEN 900
880 R = R + 1
890 KO = 5 - 2 * J
900    NEXT J
910    IF R = 0 THEN 960
920    IF R = 2 THEN 940
925    REM   ==SINGULAR POINT==
930    GOSUB 1050
935    REM =="FLAT" POINT==
940    GOSUB 1560
950    GOTO 1000
955    REM   ==REGULAR POINT==
960 KO = 0
970 KO = KO + 1
980    GOSUB 1050
990    IF KO < 3 THEN 970
```

```
995   REM   ==UPDATE ARRAY A==
1000   FOR I = 1 TO 3
1010 A(I,I) = S(I) + P
1020   NEXT I
1030   GOSUB 2200
1040   RETURN
1045   REM   ==PRINCIPAL DIRECTIONS SUBROUTINE==
1048   REM   ==STATE HOMOGENEOUS EQTS==
1050   FOR K = 1 TO 3
1060 A(K,K) = S(K) - D(KO)
1070   NEXT K
1080   GOSUB 2200
1085   REM   ==FACTORIZATION OF A==
1090   FOR K = 1 TO 2
1100 I1 = K + 1
1110   FOR I = I1 TO 3
1120   IF  ABS (A(K,K)) > E1 THEN 1230
1130   FOR L = I1 TO 3
1140   IF  ABS (A(L,K)) < E1 THEN 1210
1150   FOR J = 1 TO 3
1160 E = A(K,J)
1170 A(K,J) = A(L,J)
1180 A(L,J) = E
1190   NEXT J
1200   GOTO 1230
1210   NEXT L
1220   GOTO 1280
1230 E = A(I,K) / A(K,K)
1240   FOR J = K TO 3
1250 A(I,J) = A(I,J) - A(K,J) * E
1260   NEXT J
1270   NEXT I
1280   NEXT K
1285   REM   ==VECTOR N COMPUTATION==
1290 S = 0
1300   FOR I = 1 TO 3
1310 K = 4 - I
1320 N(K) = 0
1330   IF  ABS (A(K,K)) > E1 THEN 1360
1340 N(K) = 1
1350   GOTO 1410
1360   IF K = 3 THEN 1410
1370 J1 = K + 1
1380   FOR J = J1 TO 3
1390 N(K) = N(K) - N(J) * A(K,J) / A(K,K)
1400   NEXT J
1410 S = S + N(K) * N(K)
1420   NEXT I
1430 S =   SQR (S)
1435   REM   ==NORMALIZATION OF N==
1440   FOR K = 1 TO 3
1450 N(K) = N(K) / S
1460   NEXT K
1465   REM   ==OUTPUT MODULE REGULAR PT==
1470   PRINT
```

```
1480   PRINT "  SIGMA(";KO;") =";P(KO)
1490   FOR I = 1 TO 3
1500   PRINT
1510   PRINT "  N(";I;") =";N(I)
1520   NEXT I
1530   PRINT
1540   GOSUB 1870
1550   RETURN
1560   PRINT
1565   REM  ==OUTPUT MODULE NON-REGULAR POINT==
1570   IF R = 2 THEN 1700
1580 K1 = 1
1590   IF KO = 3 THEN 1610
1600 K1 = 3
1610   PRINT "  SIGMA(2)=SIGMA(";K1;") =";P(2)
1620   PRINT
1630   PRINT
1640   PRINT "  **FLAT POINT**"
1650   PRINT
1660   PRINT "  ANY DIRECTION ORTHOGONAL"
1670   PRINT "  TO   N   IS PRINCIPAL"
1680   PRINT
1690   RETURN
1700   PRINT "  SIGMA(1)=SIGMA(2)=SIGMA(3)= ";P(2)
1710   PRINT
1720   PRINT
1730   PRINT "  **SINGULAR POINT**"
1740   PRINT
1750   PRINT "  ANY DIRECTION IS PRINCIPAL"
1760   PRINT
1770   RETURN
1775   REM ==INVARIANTS OUTPUT==
1780   PRINT
1790   PRINT "  J1 = ";J1
1800   PRINT
1810   PRINT "  J2 = ";J2
1820   PRINT
1830   PRINT "  J3 = ";J3
1840   PRINT
1850   PRINT
1860   RETURN
1865   REM  =="SEPARATOR" ROUTINE==
1870   PRINT "  PRESS ANY KEY TO CONTINUE ";
1880   GET Z$
1890   PRINT
1900   PRINT
1910   PRINT
1920   RETURN
1930 S = 0
1940   PRINT
1950   PRINT ">INPUT DIRECTION"
1960   FOR I = 1 TO 3
1970   PRINT
1980   PRINT "  N(";I;") = ";
1990   INPUT N(I)
```

```
2000 S = S + N(I) * N(I)
2010  NEXT I
2020 S =  SQR (S)
2030  FOR I = 1 TO 3
2040 N(I) = N(I) / S
2050  NEXT I
2060  FOR I = 1 TO 3
2070 P(I) = 0
2080  FOR J = 1 TO 3
2090 P(I) = P(I) + A(I,J) * N(J)
2100  NEXT J
2110  NEXT I
2120  PRINT
2130  PRINT ">COMPONENTS ON  N  DIRECTION"
2140  FOR I = 1 TO 3
2150  PRINT
2160  PRINT "  SIG(";I;") = ";P(I)
2170  NEXT I
2180  PRINT
2190  RETURN
2200  FOR J = 1 TO 2
2210 L = J + 1
2220  FOR I = L TO 3
2230 K = I + J + 1
2240 A(I,J) = S(K)
2250 A(J,I) = S(K)
2260  NEXT I
2270  NEXT J
2280  RETURN
3000  END
```

References

Fung, Y. C. *Foundations of Solid Mechanics.* Prentice-Hall, 1965.

Katchanov. *Theory of Plasticity.* Moscow: Mir, 1974.

Sokolnikoff. *Mathematical Theory of Elasticity.* McGraw-Hill, 1956.

Sokolinkoff. *Tensor Analysis.* John Wiley & Sons, 1964.

Geometric Properties of an Arbitrary Plane Domain

This program will compute the following properties of an arbitrary plane domain:

- Area
- Position of the centroid with respect to the given axis
- First moments with respect to the given axis
- Inertia moments and inertia product with respect to the given axis
- Position of the principal axes of inertia
- Principal moments of inertia

The given domain is defined by its exterior contour and any arbitrary number of interior contours (i.e., no restriction is placed on the degree of connectivity of the domain). In turn, each contour is defined by any arbitrary number of points.

Finally, each point is defined by its coordinates with respect to an arbitrary system of Cartesian coordinates. The only restrictions are as follows:

- The points for each contour must be numbered consecutively, either clockwise or counter-clockwise.
- The program always assumes that each consecutive point is connected to its neighbors by a straight segment.

The following diagram illustrates the convention chosen to define the principal axes:

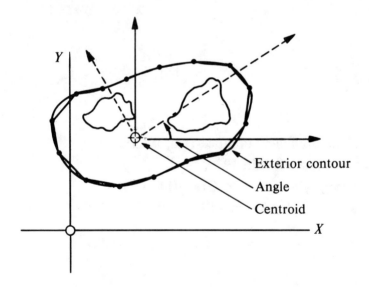

Example

Compute the geometric properties of the following domain:

```
RUN
GEOMETRIC PROPERTIES OF AN
ARBITRARY PLANE DOMAIN
```

```
TOTAL NUMBER OF CONTOURS ? 4

FOR EACH CONTOUR
VERTICES MUST BE NUMBERED CONSECUTIVELY
STARTING WITH 1


DEFINITION OF EXTERIOR CONTOUR

NUMBER OF VERTICES ? 8

ENTER VERTEX COORDINATES (X,Y)
VERTEX 1   ?7.5,3
VERTEX 2   ?18.5,3
VERTEX 3   ?21,13
VERTEX 4   ?23,13
VERTEX 5   ?23,15
VERTEX 6   ?3,15
VERTEX 7   ?3,13
VERTEX 8   ?5,13


DEFINITION OF INTERIOR CONTOUR 1

NUMBER OF VERTICES ? 4

ENTER VERTEX COORDINATES (X,Y)
VERTEX 1   ?8,4
VERTEX 2   ?10,4
VERTEX 3   ?10,13
VERTEX 4   ?6,13


DEFINITION OF INTERIOR CONTOUR 2

NUMBER OF VERTICES ? 4

ENTER VERTEX COORDINATES (X,Y)
VERTEX 1   ?11,13
VERTEX 2   ?15,13
VERTEX 3   ?15,4
VERTEX 4   ?11,4


DEFINITION OF INTERIOR CONTOUR 3

NUMBER OF VERTICES ? 4

ENTER VERTEX COORDINATES (X,Y)
VERTEX 1   ?20,13
VERTEX 2   ?16,13
VERTEX 3   ?16,4
VERTEX 4   ?18,4
```

```
GEOMETRIC PROPERTIES
WITH RESPECT TO GIVEN AXES

   AREA = 85

COORDINATES OF CENTROID
   XG = 13
   YG = 10.4666667

MOMENTS OF INERTIA
   IXX = 10716
   IYY = 16600.9583
   IXY = 11565.6667

STATIC MOMENTS
   MXX = 889.666667
   MYY = 1105

PRESS SPACE BAR TO CONTINUE


INERTIA TENSOR WITH RESPECT
TO PARALLEL AXES THRU CENTROID

   IXXG = 1404.15555
   IYYG = 2235.95833
   IXYG = 0


PRINCIPAL AXES AND MOMENTS OF INERTIA

   ANGLE = 0
   IMAX = 2235.95833
   IMIN = 1404.15555
```

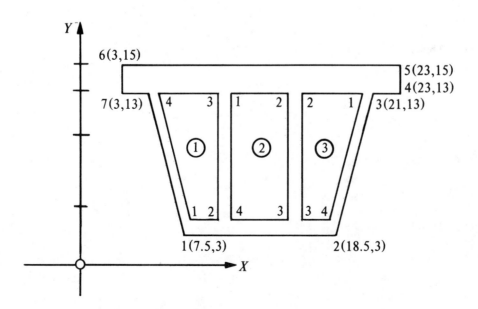

Exterior Contour	Contour 1	Contour 2	Contour 3
1(4.5,3)	1(8,4)	1(11,13)	1(20,13)
2(18.5,3)	2(10,4)	2(15,13)	2(16,13)
3(21,13)	3(10,13)	3(15,4)	3(16,4)
4(23,13)	4(6,13)	4(11,4)	4(18,4)
5(23,15)			
6(3,15)			
7(5,13)			

Program Listing

```
1    HOME
10   PRINT "GEOMETRIC PROPERTIES OF AN"
20   PRINT "ARBITRARY PLANE DOMAIN"
30   PRINT
40   PRINT
45   REM : DATA INPUT MODULE
50   PRINT
60   INPUT "TOTAL NUMBER OF CONTOURS ? ";N3
70 N3 = N3 - 1
80   PRINT
90   PRINT "FOR EACH CONTOUR"
100   PRINT "VERTICES MUST BE NUMBERED CONSECUTIVELY"
110   PRINT "STARTING WITH 1"
120   PRINT
130   PRINT
140   PRINT
150   PRINT "DEFINITION OF EXTERIOR CONTOUR"
160   PRINT
170   INPUT "NUMBER OF VERTICES ? ";N1
180   PRINT
190 N = N1
195   REM ==COMPUTE ELEMENT PROPERTIES FOR EXTERIOR CONTOUR==
200   GOSUB 1030
205   REM ==CHECK FOR AREA SIGN==
210 I0 = 1
220   IF S0 > 0 THEN 240
230 I0 =  - 1
235   REM ==SIMPLY CONNECTED DOMAIN PROPERTIES==
240 A = I0 * S0
250 Q1 = I0 * M1
260 Q2 = I0 * M2
270 I1 =   ABS (R1)
280 I2 =   ABS (R2)
290 I3 = I0 * R3
295   REM ==CHECK FOR MULTIPLE CONNECTIVITY==
300   IF N3 = 0 THEN 500
305   REM ==INTERIOR DOMAINS==
310 J = 0
320 J = J + 1
330   PRINT
340   PRINT
350   PRINT "DEFINITION OF INTERIOR CONTOUR ";J
360   PRINT
370   INPUT "NUMBER OF VERTICES ? ";N2
```

```
380 N = N2
385   REM ==COMPUTE INTERIOR ELEMENT PROPERTIES==
390   GOSUB 1030
395   REM ==CHECK FOR AREA SIGN==
400 IO = 1
410   IF SO > 0 THEN 430
420 IO =  - 1
425   REM ==UPDATE PROPERTIES==
430 A = A - IO * SO
440 Q1 = Q1 - IO * M1
450 Q2 = Q2 - IO * M2
460 I1 = I1 -  ABS (R1)
470 I2 = I2 -  ABS (R2)
480 I3 = I3 - IO * R3
485   REM ==CHECK FOR CONNECTIVITY ORDER==
490   IF J < N3 THEN 320
495   REM ==CENTROID COORDINATES==
500 X = Q2 / A
510 Y = Q1 / A
520   PRINT
530   PRINT
540   PRINT "GEOMETRIC PROPERTIES"
550   PRINT "WITH RESPECT TO GIVEN AXES"
560   PRINT
570   PRINT "  AREA = ";A
580   PRINT
590   PRINT "COORDINATES OF CENTROID"
600   PRINT "  XG = ";X
610   PRINT "  YG = ";Y
620   PRINT
630   PRINT "MOMENTS OF INERTIA"
640   PRINT "  IXX = ";I1
650   PRINT "  IYY = ";I2
660   PRINT "  IXY = ";I3
670   PRINT
680   PRINT "STATIC MOMENTS"
690   PRINT "  MXX = ";Q1
700   PRINT "  MYY = ";Q2
710   PRINT
720   PRINT "PRESS SPACE BAR TO CONTINUE"
730   GET W$
735   REM ==CENTROIDAL PROPERTIES WITH RESPECT TO PARALLEL AXES==
740 J1 = I1 - A * Y * Y
750 J2 = I2 - A * X * X
760 J3 = I3 - A * X * Y
770 C1 = (J1 + J2) / 2
780 C2 = (J2 - J1) / 2
790 C3 =  SQR (C2 * C2 + J3 * J3)
795   REM ==PRINCIPAL PROPERTIES==
800 P1 = C1 + C3
810 P2 = C1 - C3
820 PO = 3.141592653589
825   REM ==COMPUTE ORIENTATION OF PRINCIPAL AXES==
830 T = PO / 4
840   IF ( ABS (C2) < .00001) THEN 860
```

```
850 T =  ATN (J3 / C2) / 2
860 T = T * 180 / PO
870   PRINT
880   PRINT
890   PRINT "INERTIA TENSOR WITH RESPECT"
900   PRINT "TO PARALLEL AXES THRU CENTROID"
910   PRINT
920   PRINT "   IXXG = ";J1
930   PRINT "   IYYG = ";J2
940   PRINT "   IXYG = ";J3
950   PRINT
960   PRINT
970   PRINT "PRINCIPAL AXES AND MOMENTS OF INERTIA"
980   PRINT
990   PRINT "   ANGLE = ";T
1000   PRINT "   IMAX = ";P1
1010   PRINT "   IMIN = ";P2
1020   GOTO 1400
1025   REM ==INITIALIZE ELEMENT VARIABLES==
1030 S0 = 0
1040 M1 = 0
1050 M2 = 0
1060 R1 = 0
1070 R2 = 0
1080 R3 = 0
1085   REM ==INPUT COORDINATES==
1090   PRINT
1100   PRINT "ENTER VERTEX COORDINATES (X,Y)"
1110   INPUT "VERTEX 1   ?";X1,Y1
1120 X0 = X1
1130 Y0 = Y1
1140 I = 2
1150   PRINT "VERTEX ";I; SPC( 2);
1160   INPUT X2,Y2
1165   REM ==ELEMENT AREA==
1170 S = (X1 * Y2 - X2 * Y1) / 2
1175   REM ==ELEMENT CENTROID==
1180 Z1 = (X1 + X2) / 3
1190 Z2 = (Y1 + Y2) / 3
1195   REM ==FIRST MOMENTS==
1200 M1 = M1 + S * (Z2)
1210 M2 = M2 + S * (Z1)
1215   REM ==INERTIA MOMENTS==
1220 R1 = R1 + S * (Y1 * Y1 + Y2 * Y2 + Y1 * Y2) / 6
1230 R2 = R2 + S * (X1 * X1 + X2 * X2 + X1 * X2) / 6
1240 R3 = R3 + S * (X1 * Y1 + X2 * Y2 + (X1 * Y2 + X2 * Y1) / 2) / 6
1245   REM ==UPDATE AREA==
1250 S0 = S0 + S
1260 X1 = X2
1270 Y1 = Y2
1280 I = I + 1
1285   REM ==CHECK FOR LAST VERTEX==
1290   IF I < N + 1 THEN 1150
1300   IF I > N + 1 THEN 1340
1305   REM ==LAST ELEMENT==
```

```
1310  X2 = X0
1320  Y2 = Y0
1330   GOTO 1170
1340   RETURN
1400   END
```

References

Popov. *Mechanics of Materials.* Prentice-Hall, 1976.

Bending Moments and Shear Force Envelopes

This program will compute the envelope of bending moment and shear force of a beam segment subjected to various load conditions, as shown is the illustration below:

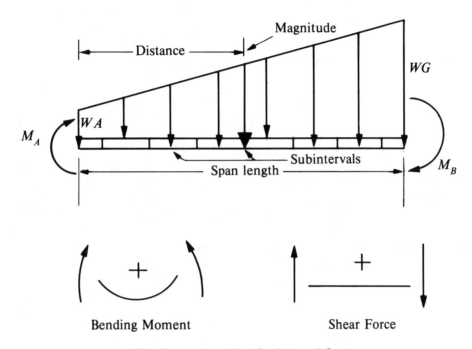

Bending Moment   Shear Force

Bending convention for internal force

Program Notes

The maximum number of output points for this program has been set to 25. To modify this, change lines 10 and 20 as follows:

```
10 DIM X(N), V(N,2N), M(N,2)
20 DIM S(N), B(N)
```

where N is the maximum desired.

Example

Given the beams below, compute the envelopes of bending moments and shear forces along the left spans for load multipliers of 1.0 in the first case and 1.25 in the second:

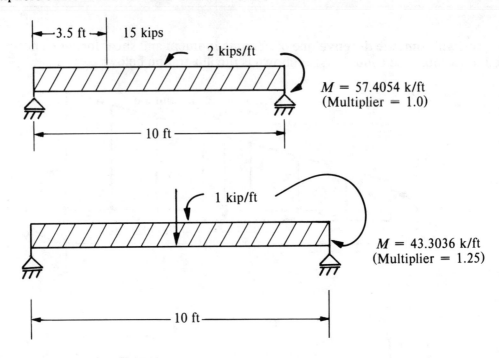

```
RUN
BENDING MOMENT AND SHEAR FORCE ENVELOPES

**PROBLEM SPECIFICATION**

SPAN LENGTH = 10

NUMBER OF LOAD CASES = 2

**OUTPUT DEFINITION**

NUMBER OF SUBINTERVALS = 10

NUMBER OF ADDITIONAL

OUTPUT POINTS = 1

ENTER ADD. POINTS POSITION

POINT NO. 1
X = 3.5

**LOAD CASES SPECIFICATION**

> LOAD CASE 1

MULTIPLIER = 1

END MOMENTS
```

```
MA = 0
MB = 57.4054

DISTRIBUTED LOAD

WA = 2
WB = 2

NUMBER OF POINT LOADS = 1

LOAD 1

MAGNITUDE = 15
DISTANCE = 3.5

> LOAD CASE 2

MULTIPLIER = 1.25

END MOMENTS

MA = 0
MB = 43.3036

DISTRIBUTED LOAD

WA = 1
WB = 1

NUMBER OF POINT LOADS = 0

BENDING MOMENT ENVELOPE
************************


POINT 1  AT X = 0

MAX M = 0
MIN M = 0

POINT 2  AT X = 1

MAX M = 13.00946
MIN M = 0

POINT 3  AT X = 2

MAX M = 24.01892
MIN M = -.825899998

POINT 4  AT X = 3

MAX M = 33.02838
MIN M = -3.11385

POINT 5  AT X = 3.5
```

```
MAX M = 36.78311
MIN M = -4.726575

POINT 6  AT X = 4

MAX M = 32.53784
MIN M = -6.6518

POINT 7  AT X = 5

MAX M = 22.5473
MIN M = -11.43975

POINT 8  AT X = 6

MAX M = 10.55676
MIN M = -17.4777

POINT 9  AT X = 7

MAX M = 0
MIN M = -24.76565

POINT 10  AT X = 8

MAX M = 0
MIN M = -33.3036

POINT 11  AT X = 9

MAX M = 0
MIN M = -43.09155

POINT 12  AT X = 10

MAX M = 0
MIN M = -57.4054

IS IT OK TO CONTINUE (Y/N) ?



SHEAR FORCE ENVELOPE
*********************


POINT 1  AT X = 0

MAX V = 14.00946
MIN V = 0

POINT 2  AT X = 1

MAX V = 12.00946
MIN V = -.412949999
```

POINT 3   AT X = 2

MAX V = 10.00946
MIN V = -1.66295

POINT 4   AT X = 3

MAX V = 8.00946001
MIN V = -2.91295

POINT 5   AT X = 3.5

MAX V = 0
MIN V = -7.99054

POINT 6   AT X = 4

MAX V = 0
MIN V = -8.99054

POINT 7   AT X = 5

MAX V = 0
MIN V = -10.99054

POINT 8   AT X = 6

MAX V = 0
MIN V = -12.99054

POINT 9   AT X = 7

MAX V = 0
MIN V = -14.99054

POINT 10   AT X = 8

MAX V = 0
MIN V = -16.99054

POINT 11   AT X = 9

MAX V = 0
MIN V = -18.99054

POINT 12   AT X = 10

MAX V = 0
MIN V = -20.99054

Program Listing

```
0   HOME
5   PRINT "BENDING MOMENT AND SHEAR FORCE ENVELOPES"
```

```
10    DIM X(25),V(25,2),M(25,2)
20    DIM S(25),B(25)
25    REM ==INPUT GENERAL DATA AND INITIALIZE==
30    GOSUB 80
35    REM  ==DEFINE OUTPUT POINTS==
40    GOSUB 430
45    REM ==ENVELOPE COMPUTATION==
50    GOSUB 690
55    REM ==OUTPUT RESULTS==
60    GOSUB 1850
70    GOTO 2200
80    PRINT
140    PRINT
160    PRINT
170    PRINT "**PROBLEM SPECIFICATION**"
180    PRINT
190    INPUT "SPAN LENGTH = ";L
200    PRINT
210    INPUT "NUMBER OF LOAD CASES = ";N
220    PRINT
230    PRINT "**OUTPUT DEFINITION**"
240    PRINT
250    INPUT "NUMBER OF SUBINTERVALS = ";K1
260    PRINT
270    PRINT "NUMBER OF ADDITIONAL"
280    PRINT
290    INPUT "OUTPUT POINTS = ";K2
300    PRINT
310 T = K1 + K2 + 1
320    REM ==INITIALIZE VARIABLES==
330    FOR I = 1 TO T
340 X(I) = 0
350    FOR J = 1 TO 2
360 M(I,J) = 0
370 V(I,J) = 0
380    NEXT J
390    NEXT I
400    RETURN
410    REM ==DEFINE AND ORDER OUTPUT POINTS==
430 D = L / (K1)
440 K3 = K1 + 1
450    FOR I = 1 TO K3
460 X(I) = (I - 1) * D
470    NEXT I
480 K4 = K3 + 1
490    IF K2 = 0 THEN 670
500    PRINT "ENTER ADD. POINTS POSITION"
510    FOR I = K4 TO T
520 I1 = I - K3
530    PRINT
540    PRINT "POINT NO. ";I1
550    INPUT "X = ";X(I)
560    NEXT I
570 T1 = T - 1
580    FOR I = 1 TO T1
590 I1 = I + 1
```

```
600    FOR J = I1 TO T
610    IF X(J) >  = X(I) THEN 650
620 Y = X(I)
630 X(I) = X(J)
640 X(J) = Y
650    NEXT J
660    NEXT I
670    RETURN
680    REM ==ENVELOPE DETERMINATION==
690    PRINT
700    PRINT "**LOAD CASES SPECIFICATION**"
710    PRINT
720 K = 0
730 K = K + 1
740    IF K > N THEN 1180
750    PRINT
760    REM ==INITIALIZE==
770    FOR I = 1 TO T
780 S(I) = 0
790 B(I) = 0
800    NEXT I
810    PRINT "> LOAD CASE ";K
820    PRINT
830    INPUT "MULTIPLIER = ";Z
840    PRINT
850    PRINT "END MOMENTS"
860    PRINT
870    INPUT "MA = ";MO
880    INPUT "MB = ";M1
890    REM ==EFFECT OF END MOMENTS==
900    GOSUB 1200
910    PRINT
920    PRINT "DISTRIBUTED LOAD"
930    PRINT
940    INPUT "WA = ";WO
950    INPUT "WB = ";W1
960    REM ==EFFECT OF DISTR. LOAD==
970    GOSUB 1310
980    PRINT
990    INPUT "NUMBER OF POINT LOADS = ";N3
1000   IF N3 = 0 THEN 1150
1010 J = 0
1020 J = J + 1
1030   IF J > N3 THEN 1150
1040   PRINT
1050   PRINT "LOAD ";J
1060   PRINT
1070   INPUT "MAGNITUDE = ";P
1080   INPUT "DISTANCE = ";A1
1090   REM ==EFFECT OF POINT LOAD==
1100   GOSUB 1460
1110   REM  ==NEXT POINT LOAD==
1120   GOTO 1020
1130   REM ==CHECK FOR MAXIMUM AMPLITUDES==
1150   GOSUB 1660
```

```
1160   REM ==NEXT LOAD CASE==
1170   GOTO 730
1180   RETURN
1190   REM ==EFFECT OF END MOMENTS==
1200 VO = (MO + M1) / L
1210 I = 0
1220 I = I + 1
1230   IF I > T THEN 1290
1240 S1 =   - VO
1250 B1 = MO + S1 * X(I)
1260 S(I) = S(I) + S1
1270 B(I) = B(I) + B1
1280   GOTO 1220
1290   RETURN
1300   REM ==EFFECT OF DISTR. LOAD==
1310 VO = (WO / 3 + W1 / 6) * L
1320 A1 = (W1 - WO) / L
1330 I = 0
1340 I = I + 1
1350   IF I > T THEN 1440
1360 X1 = X(I)
1370 X2 = X1 * X1
1380 X3 = X2 * X1
1390 S2 = VO - WO * X1 - A1 * X2 / 2
1400 B2 = VO * X1 - WO * X2 / 2 - A1 * X3 / 6
1410 S(I) = S(I) + S2
1420 B(I) = B(I) + B2
1430   GOTO 1340
1440   RETURN
1450   REM ==EFFECT OF POINT LOAD==
1460 C = L - A1
1470 VO = P * C / L
1480 V1 = P * A1 / L
1490 I = 0
1500 I = I + 1
1510   IF I > T THEN 1640
1520 X1 = X(I)
1530   IF X1 >  = A1 THEN 1590
1540 S3 = VO
1550 B3 = (VO * X1)
1560 S(I) = S(I) + S3
1570 B(I) = B(I) + B3
1580   GOTO 1500
1590 S3 =   - V1
1600 B3 = V1 * (L - X1)
1610 S(I) = S(I) + S3
1620 B(I) = B(I) + B3
1630   GOTO 1500
1640   RETURN
1650   REM ==ENVELOPE GENERATION==
1660   FOR I = 1 TO T
1670   REM ==APPLY MULTIPLIER==
1680 S1 = S(I) * Z
1690 B1 = B(I) * Z
1700   REM   ==MAX SHEAR+==
```

```
1710   IF S1 <  = V(I,1) THEN 1740
1720 V(I,1) = S1
1730   REM ==MAX SHEAR-==
1740   IF S1 >  = V(I,2) THEN 1770
1750 V(I,2) = S1
1760   REM ==MAX MOMENT+==
1770   IF B1 <  = M(I,1) THEN 1800
1780 M(I,1) = B1
1790   REM ==MAX MOMENT-==
1800   IF B1 >  = M(I,2) THEN 1820
1810 M(I,2) = B1
1820   NEXT I
1830   REM ==OUTPUT RESULTS==
1840   RETURN
1850   PRINT
1860   PRINT "BENDING MOMENT ENVELOPE"
1870   PRINT "***********************"
1880   PRINT
1890   FOR I = 1 TO T
1900   PRINT
1910   PRINT "POINT ";I; SPC( 2);"AT X = ";X(I)
1920   PRINT
1930   PRINT "MAX M = ";M(I,1)
1940   PRINT "MIN M = ";M(I,2)
1950   NEXT I
1960   PRINT
1970   PRINT "IS IT OK TO CONTINUE (Y/N) ";
1980   INPUT C$
1990   IF C$ = "N" THEN 2120
2000   PRINT
2010   PRINT
2020   PRINT "SHEAR FORCE ENVELOPE"
2030   PRINT "********************"
2040   PRINT
2050   FOR I = 1 TO T
2060   PRINT
2070   PRINT "POINT ";I; SPC( 2);"AT X = ";X(I)
2080   PRINT
2090   PRINT "MAX V = ";V(I,1)
2100   PRINT "MIN V = ";V(I,2)
2110   NEXT I
2120   RETURN
2200   END
```

Fixed End Moments and Stiffness Coefficients for Beams

This program computes the fixed end forces and stiffness properties for elastic straight beams, subjected to arbitrary loading.

In order to use this program you will need to idealize your data into a series of segments such that:

- The cross-section moment of inertia varies approximately linearly
- The distributed load on the segments varies approximately linearly
- There are no concentrated loads within each segment span

Program Notes

The program requires as inputs:

- The modulus of elasticity
- The total length of the beam
- The number of segments
- Segment data (either the moment of inertia or the depth of the end cross sections)

The maximum number of segments the program can handle is set at 20, but this can be altered by changing the DIM statement at line 10 as follows:

$$10 \ DIM \ F1(N), \ F2(N), \ F3(N), \ Z(N)$$

where N is the maximum number of segments.

Example

Compute the fixed-end moments and stiffness coefficients for the beam shown in the illustration below:

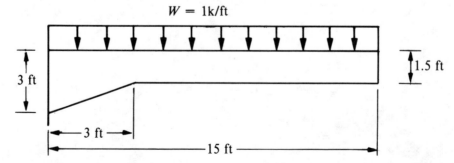

RUN
FIXED END MOMENTS AND STIFFNESS

  COEFFICIENTS FOR BEAMS

ENTER GENERAL DATA

ELASTIC MODULUS = 1.0

TOTAL LENGTH = 15.0

NUMBER OF SEGMENTS = 2

CROSS SECTION GEOMETRY
SPECIFICATION TYPES :

TYPE 1 : MOMENT OF INERTIA
TYPE 2 : DEPTH

TYPE = 2

ENTER SEGMENT DATA

SEGMENT 1

LENGTH = 3.0

END SECTIONS DEPTHS
DO = 3.0
D1 = 1.5

DISTRIBUTED LOAD
WO = 1.0
W1 = 1.0

NODAL LOADS
FORCE = 0
MOMENT = 0

SEGMENT 2

LENGTH = 12.0

END SECTIONS DEPTHS
DO = 1.5
D1 = 1.5

DISTRIBUTED LOAD
WO = 1.0
W1 = 1.0

NODAL LOADS
FORCE = 0
MOMENT = 0

SOLUTION
****

FIXED END FORCES

VA = 8.3807774
MA = -27.9172494

VB = -6.61922261
MB = 14.7055885

STIFFNESS COEFFICIENTS

KAA = .131786862
KBB = .0855994079
KAB = .061933016

Program Listing

```
0    HOME
1    PRINT "FIXED END FORCES AND STIFFNESS "
2    PRINT "COEFFICIENTS FOR STRAIGHT ELASTIC BEAMS"
10   DIM F1(20),F2(20),F3(20),Z(20)
20   GOSUB 80
25   REM ==INPUT SEGMENT DATA==
30   GOSUB 360
35   REM ==COMPUTE FIXED END FORCES==
40   GOSUB 1340
45   REM ==COMPUTE STIFFNESS COEFFICIENTS==
50   GOSUB 1540
60   GOTO 1950
80   PRINT
100  PRINT
160  PRINT
170  PRINT "ENTER GENERAL DATA"
180  PRINT
190  INPUT "ELASTIC MODULUS = ";E
200  PRINT
210  INPUT "TOTAL LENGTH = ";L9
220  PRINT
230  INPUT "NUMBER OF SEGMENTS = ";N1
240  PRINT
250  PRINT "CROSS SECTION GEOMETRY"
260  PRINT "SPECIFICATION TYPES :"
270  PRINT
280  PRINT "TYPE 1 : MOMENT OF INERTIA"
290  PRINT "TYPE 2 : DEPTH"
300  PRINT
310  INPUT "TYPE = ";TO
320  PRINT
330  PRINT "ENTER SEGMENT DATA"
340  PRINT
350  RETURN
360  N = N1 + 1
370  G1 = 0
380  G2 = 0
390  G3 = 0
400  D1 = 0
```

```
410 D2 = 0
420 X1 = 0
430 X2 = 0
440 Z0 = 0
450 J = 0
460 J = J + 1
470 J1 = J - 1
480   IF J >  = N THEN 1330
485   REM ==INPUT SEGMENT DATA==
490   PRINT
500   PRINT "SEGMENT ";J
510   PRINT
520   INPUT "LENGTH = ";L
530   PRINT
540   IF T0 > 1 THEN 590
550   PRINT "END SECTIONS MOMENTS OF INERTIA"
560   INPUT "I0 = ";I0
570   INPUT "I1 = ";I1
580   GOTO 640
590   PRINT "END SECTIONS DEPTHS"
600   INPUT "D0 = ";E0
610   INPUT "D1 = ";E1
620 I0 = E0 * E0 * E0 / 12
630 I1 = E1 * E1 * E1 / 12
640   PRINT
650   PRINT "DISTRIBUTED LOAD"
660   INPUT "W0 = ";W0
670   INPUT "W1 = ";W1
680   PRINT
690   PRINT "NODAL LOADS"
700   INPUT "FORCE = ";P0
710   INPUT "MOMENT = ";M0
720   PRINT
725   REM ==CHECK FOR INPUT ERROR==
730 Z0 = Z0 + L
740   IF Z0 < 1.001 * L9 THEN 770
750   PRINT "***WARNING***"
760   PRINT "TOTAL LENGTH EXCEEDED"
770   PRINT
800   GOTO 820
810   GOTO 490
815   REM ==COMPUTE FLEXIBILITIES==
820   GOSUB 1650
830 F1(J) = H2
840 F2(J) = H0
850 F3(J) = H1
860 Z(J) = Z0
870 G9 = L9 - Z(J)
880 G1 = G1 + H2 + 2 * G9 * H1 + G9 * G9 * H0
890 G2 = G2 + H0
900 G3 = G3 + H1 + G9 * H0
905   REM ==CHECK FOR DISTR. LOAD==
910   IF W0 <  > 0 THEN 940
920   IF W1 <  > 0 THEN 940
930   GOTO 1040
```

```
935    REM ==EFFECT OF DISTRIBUTED LOAD==
940 A1 = (W1 - W0) / L
950 A2 = W1 / 2
960 A6 = A1 / 6
965    REM ==COMPUTE INTEGRALS==
970    GOSUB 1840
975    REM ==COMPUTE AND ADD DEFLECTIONS AT TIP==
980 V1 = A2 * H3 - A6 * H4
990 V2 = A2 * H2 - A6 * H3
1000 D1 = D1 + V1 + G9 * V2
1010 D2 = D2 + V2
1015    REM ==MODIFY NODAL LOADS==
1020 P0 = P0 + (W0 + W1) * L / 2
1030 M0 = M0 + (W0 + 2 * W1) * L * L / 6
1035    REM ==CHECK FOR NODAL LOADS==
1040    IF P0 <  > 0 THEN 1070
1050    IF M0 <  > 0 THEN 1070
1060    GOTO 460
1065    REM ==LEFT SUPPORT REACTIONS==
1070    IF J > 1 THEN 1110
1080 X1 = X1 + P0
1090 X2 = X2 + M0
1100    GOTO 460
1110 X1 = X1 + P0
1120 X2 = X2 + M0 + P0 * Z(J1)
1130 C1 = 0
1140 C2 = 0
1150 K = 0
1160 K = K + 1
1170    IF K >  = J THEN 1300
1180 S1 = P0
1190 L2 = Z(J1) - Z(K)
1200 S2 = L2 * P0 + M0
1210 G8 = L9 - Z(K)
1220 B1 = F1(K)
1230 B2 = F2(K)
1240 B3 = F3(K)
1250 A1 = B1 + G8 * B3
1260 A2 = B3 + G8 * B2
1270 C1 = C1 + A1 * S1 + A2 * S2
1280 C2 = C2 + B3 * S1 + B2 * S2
1290    GOTO 1160
1295    REM ==ACCUMULATE DEFLECTION DUE TO NODAL LOADS==
1300 D1 = D1 + C1
1310 D2 = D2 + C2
1320    GOTO 460
1330    RETURN
1335    REM ==COMPUTE FIXED END LOADS STIFFNESS MATRIX==
1340 D9 = G1 * G2 - G3 * G3
1350 S1 = G2 / D9
1360 S2 = G1 / D9
1370 S3 =  - G3 / D9
1380 X3 =  - (S1 * D1 + S3 * D2)
1390 X4 =  - (S3 * D1 + S2 * D2)
1400 X1 = X1 + X3
```

```
1410 X2 =   - (X2 + X4 + X3 * L9)
1415  REM ==OUTPUT==
1420  PRINT
1430  PRINT "SOLUTION"
1440  PRINT "********"
1450  PRINT
1460  PRINT "FIXED END FORCES"
1470  PRINT
1480  PRINT "VA = ";X1
1490  PRINT "MA = ";X2
1500  PRINT
1510  PRINT "VB = ";X3
1520  PRINT "MB = ";X4
1530  RETURN
1535  REM ==COMPUTATION OF STIFFNESS COEFGFICIENTS COORDINATE TRANSFORMATION=
1540 L2 = L9 * L9
1550 K1 = S1 * L2 + 2 * S3 * L9 + S2
1560 K2 = S2
1570 K3 =   - (S3 * L9 + S2)
1575  REM ==OUTPUT==
1580  PRINT
1590  PRINT "STIFFNESS COEFFICIENTS"
1600  PRINT
1610  PRINT "KAA = ";K1
1620  PRINT "KBB = ";K2
1630  PRINT "KAB = ";K3
1640  RETURN
1645  REM ==INTEGRALS FOR FLEXIBILITY COMPUTATION==
1646  REM  FLEXIBILITY COMPUTATION
1650  IF IO = I1 THEN 1760
1660 B = (I1 - IO) / L
1665  REM ==VARIABLE INERTIA==
1670 B0 = 1 / (B * E)
1680 B1 = I1 / IO
1690 B2 = I1 / B
1700 B3 = B0 * L
1710 H0 = B0 *  LOG (B1)
1720 H1 = B2 * H0 - B3
1730 B3 = B3 * L
1740 H2 = B2 * H1 - B3 / 2
1750  RETURN
1755  REM ==UNIFORM CROSS-SECTION==
1760 E0 = 1 / (E * IO)
1770 B4 = L
1780 H0 = E0 * B4
1790 B4 = B4 * L
1800 H1 = E0 * B4 / 2
1810 B4 = B4 * L
1820 H2 = E0 * B4 / 3
1830  RETURN
1835  REM ==INTEGRALS FOR DISTRIBUTED LOAD EFFECTYS==
1840  IF I1 = IO THEN 1900
1845  REM ==VARIABLE INERTIA==
1850 B3 = B3 * L
1860 H3 = B2 * H2 - B3 / 3
```

```
1870 B3 = B3 * L
1880 H4 = B2 * H3 - B3 / 4
1890  RETURN
1895  REM ==UNIFORM CROSS-SECTION==
1900 B4 = B4 * L
1910 H3 = E0 * B4 / 4
1920 B4 = B4 * L
1930 H4 = E0 * B4 / 5
1940  RETURN
1950  END
```

References

Clough, R., and Penzien, J. *Dynamics of Structures.* McGraw-Hill, 1971.

Livisley, R. K. *Matrix Methods of Structural Analysis.* Oxford: Pergamon Press, 1970.

Prezemieniekcki, I. G. *Theory of Matrix Structural Analysis.* McGraw-Hill, 1968.

Joint Rotations and End Moments for a Continuous Beam

This program will compute the joint rotations and the member end-moments of a continuous beam subjected to general loading.

The various program options include:

1. Column specifications
   - Number of columns
   - Columns of the same height
   - Columns of differing heights
2. Cross section elements
   - Constant cross section elements
   - Variable cross section elements

For variable cross section beams, the stiffness matrix should be such that:

$$\begin{Bmatrix} M_A \\ M_B \end{Bmatrix} = \begin{bmatrix} K_{AA} & K_{AB} \\ K_{BB} & K_{BB} \end{bmatrix} \begin{Bmatrix} 0_A \\ 0_B \end{Bmatrix}$$

In the case of variable cross section columns, the 'A' member end corresponds to that connected to the beam joint.

Program Notes

This program is designed to handle systems with up to 10 joints. This can be modified by changing the DIM statements (lines 90, 100, and 110) as follows:

```
90 DIM L(N), K1(N), K2(N), K3(N)
100 DIM Q1(N), Q2(N)
110 DIM A(N), B(N), C(N), D(N)
```

where N is the maximum number of joints.

Example

Compute the bending moment at point 2 of the continuous beam for the two load cases shown:

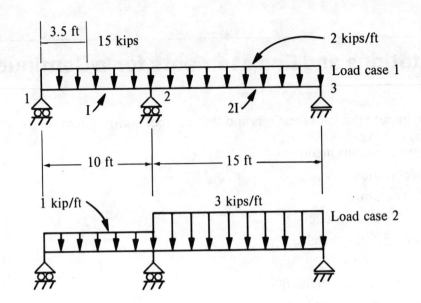

```
RUN

JOINT ROTATIONS AND END MOMENTS

 FOR A CONTINUOUS BEAM


***STRUCTURE DEFINITION***

NUMBER OF SPANS = 2

TYPE OF STRUCTURE

     1 : NO COLUMNS

     2 : EQUAL HEIGHT COLS.

     3 : DIFF'T HEIGHT COLS.

TYPE = 1

MEMBER TYPE

     1 : CONSTANT CROSS SECT.

     2 : VARIABLE CROSS SECT.

TYPE = 1

ELASTIC MODULUS = 1


***BEAM GEOMETRY***

> SPAN 1

LENGTH = 10
```

```
MOMENT OF INERTIA = 1

> SPAN 2

LENGTH = 15

MOMENT OF INERTIA = 2


***LOAD SPECIFICATION***
        LOAD CASE 1

ANY JOINT LOADS (Y/N) ?N

ANY SPAN LOADS (Y/N) ?Y

***SPAN LOADS***

> SPAN 1

DISTRIBUTED LOAD = 1

NUMBER OF POINT LOADS = 0

> SPAN 2

DISTRIBUTED LOAD = 3

NUMBER OF POINT LOADS = 0


SOLUTION FOR LOAD CASE 1
*************************

***JOINT ROTATIONS***

JOINT 1  ROT = -30.5059524

JOINT 2  ROT = 102.678571

JOINT 3  ROT = -156.808036


CONTINUE   (Y/N) ?Y


***MEMBER FORCES***

> SPAN 1
```

```
LEFT MOMENT = 0

RIGHT MOMENT = 43.3035714

> SPAN 2

LEFT MOMENT = -43.3035715

RIGHT MOMENT = 0


MORE LOAD CASES ? (Y/N) N


***ANALYSIS COMPLETE***
```

Program Listing

```
5    HOME
15   PRINT : PRINT "JOINT ROTATIONS AND END MOMENTS FOR A  CONTINUOUS BEAM"
17   GOSUB 90
20   GOSUB 560
25   REM ==TRIANGULARIZE STIFFNESS MATRIX==
30   GOSUB 1480
35   REM ==PERFORM ANALYSIS==
40   GOSUB 1540
50   PRINT
60   PRINT
70   PRINT "***ANALYSIS COMPLETE***"
80   GOTO 2880
86   REM ==INITIALIZE==
90   DIM L(10),K1(10),K2(10),K3(10)
100   DIM Q1(10),Q2(10)
110   DIM A(10),B(10),C(10),D(10)
120   GOSUB 470
125   REM ==INPUT DATA==
190   PRINT
200   PRINT
210   PRINT "***STRUCTURE DEFINITION***"
220   PRINT
230   INPUT "NUMBER OF SPANS = ";N1
240 N = N1 + 1
250   PRINT
260   PRINT "TYPE OF STRUCTURE"
270   PRINT
280   PRINT "     1 : NO COLUMNS"
290   PRINT
300   PRINT "     2 : EQUAL HEIGTH COLS."
310   PRINT
320   PRINT "     3 : DIFF'T HEIGTH COLS."
330   PRINT
340   INPUT "TYPE = ";T9
350   PRINT
360   PRINT "MEMBER TYPE"
370   PRINT
```

```
380    PRINT "        1 : CONSTANT CROSS SECT."
390    PRINT
400    PRINT "        2 : VARIABLE CROSS SECT."
410    PRINT
420    INPUT "TYPE = ";T8
430    PRINT
440    INPUT "ELASTIC MODULUS = ";E
450    PRINT
460    RETURN
470    FOR I = 1 TO N
480  A(I) = 0
490  B(I) = 0
500  C(I) = 0
510  D(I) = 0
520  Q1(I) = 0
530  Q2(I) = 0
540    NEXT I
550    RETURN
555    REM ==STIFFNESS MATRIX AND ASSEMBLY ROUTINE==
560    PRINT
570    PRINT "***BEAM GEOMETRY***"
580  I = 0
590  I = I + 1
600    IF I > N1 THEN 810
610    PRINT
620    PRINT "> SPAN ";I
630    PRINT
640    INPUT "LENGTH = ";L(I)
650    ON T8 GOTO 660,720
660    PRINT
670    INPUT "MOMENT OF INERTIA = ";J
680    PRINT
690    GOSUB 1310
700    GOSUB 1360
710    GOTO 590
720    PRINT
730    PRINT "STIFFNESS COEFFICIENTS"
740    PRINT
750    INPUT "KAA = ";K1(I)
760    INPUT "KBB = ";K2(I)
770    INPUT "KAB = ";K3(I)
780    PRINT
790    GOSUB 1360
800    GOTO 590
810    IF T9 = 1 THEN 1300
820    PRINT
830    PRINT
840    PRINT "***COLUMNS GEOMETRY***"
850    IF T9 = 3 THEN 910
860    PRINT
870    INPUT "UPPER COLS. HEIGHT = ";L1
880    PRINT
890    INPUT "LOWER COLS. HEIGHT = ";L2
900    PRINT
910  I = 0
```

```
920 I = I + 1
930   IF I > N THEN 1300
940   PRINT
950   PRINT
960   PRINT "> COLS. AT JOINT ";I
970   IF T9 > 2 THEN 1060
980   ON T8 GOTO 990,1220
990   PRINT
1000  INPUT "UPPER COL. I = ";J1
1010  PRINT
1020  INPUT "LOWER COL. I = ";J2
1030  GOSUB 1420
1040  GOSUB 1460
1050  GOTO 920
1060  ON T8 GOTO 1070,1220
1070  PRINT
1080  PRINT "UPPER COLUMN"
1090  PRINT
1100  INPUT "HEIGHT = ";L1
1110  PRINT
1120  INPUT "M. OF INERTIA = ";J1
1130  PRINT
1140  PRINT "LOWER COLUMN"
1150  PRINT
1160  INPUT "HEIGHT = ";L2
1170  PRINT
1180  INPUT "M. OF INERTIA = ";J2
1185  REM ==STIFFNESS==
1190  GOSUB 1420
1200  GOSUB 1460
1205  REM ==NEXT NODE==
1210  GOTO 920
1220  PRINT
1230  PRINT "STIFFNESS COEFFICIENTS"
1240  PRINT
1250  INPUT "UPPER COL. KAA = ";C1
1260  PRINT
1270  INPUT "LOWER COL. KAA = ";C2
1280  GOSUB 1460
1285  REM ==NEXT NODE==
1290  GOTO 920
1300  RETURN
1305  REM ==BEAM ELEMENT STIFFNESS==
1310 S = 2 * E * J / L(I)
1320 K1(I) = 2 * S
1330 K2(I) = 2 * S
1340 K3(I) = S
1350  RETURN
1360 I1 = I + 1
1370 D(I) = D(I) + K1(I)
1380 D(I1) = D(I1) + K2(I)
1390 C(I) = K3(I)
1400 A(I1) = K3(I)
1410  RETURN
1415  REM ==COLUMN STIFFNESS COEFFICIENTS==
```

```
1420 E4 = 4 * E
1430 C1 = E4 * J1 / L1
1440 C2 = E4 * J2 / L2
1450  RETURN
1460 D(I) = D(I) + C1 + C2
1470  RETURN
1475  REM ==FACTORIZATION OF STIFFNESS MATRIX==
1480  FOR I = 2 TO N
1490 I1 = I - 1
1500 A(I) = A(I) / D(I1)
1510 D(I) = D(I) - A(I) * C(I1)
1520  NEXT I
1530  RETURN
1540 K = 0
1550 K = K + 1
1560  PRINT
1570  PRINT
1580  PRINT "***LOAD SPECIFICATION***"
1590  PRINT
1600  PRINT  SPC( 6);"LOAD CASE ";K
1610  PRINT
1620  PRINT
1630  PRINT "ANY JOINT LOADS (Y/N) ";
1640  INPUT C$
1650  IF C$ = "N" THEN 1750
1660  PRINT
1670  PRINT "***JOINT LOADS***"
1680  PRINT
1690  FOR I = 1 TO N
1700  PRINT "> JOINT ";I
1710  PRINT
1720  INPUT "MOMENT = ";B(I)
1730  PRINT
1740  NEXT I
1750  PRINT
1760  PRINT "ANY SPAN LOADS (Y/N) ";
1770  INPUT C$
1780  IF C$ = "N" THEN 2090
1790  PRINT
1800  PRINT "***SPAN LOADS***"
1810 I = 0
1820 I = I + 1
1830  IF I > N1 THEN 2090
1840  PRINT
1850  PRINT "> SPAN ";I
1860  PRINT
1870  INPUT "DISTRIBUTED LOAD = ";W
1880  PRINT
1890  GOSUB 2220
1900 Q1(I) = Q1(I) + M1
1910 Q2(I) = Q2(I) + M2
1920  GOSUB 2320
1930  INPUT "NUMBER OF POINT LOADS = ";K2
1940  IF K2 = 0 THEN 2080
1950 K0 = 0
```

```
1960 K0 = K0 + 1
1970   IF K0 > K2 THEN 2080
1980   PRINT
1990   PRINT "LOAD ";K0
2000   PRINT
2010   INPUT "MAGNITUDE = ";P
2020   INPUT "DISTANCE = ";X
2030   GOSUB 2260
2040 Q1(I) = Q1(I) + M1
2050 Q2(I) = Q2(I) + M2
2060   GOSUB 2320
2065   REM ==NEXT POINT LOAD==
2070   GOTO 1960
2075   REM ==NEXT SPAN LOAD==
2080   GOTO 1820
2090   GOSUB 2360
2100   GOSUB 2460
2110   PRINT
2120   PRINT
2130   INPUT "MORE LOAD CASES ? (Y/N) ";C$
2140   IF C$ = "N" THEN 2210
2150   FOR I = 1 TO N
2160 B(I) = 0
2170 Q1(I) = 0
2180 Q2(I) = 0
2190   NEXT I
2195   REM ==INPUT LOAD DATA==
2200   GOTO 1550
2210   RETURN
2220 L0 = L(I)
2230 M2 = W * L0 * L0 / 12
2240 M1 =   - M2
2250   RETURN
2260 L0 = L(I)
2270 Y = L0 - X
2280 P1 = P * X * Y / (L0 * L0)
2290 M1 =   - P1 * Y
2300 M2 = P1 * Y
2310   RETURN
2315   REM ==FORM LOAD VECTOR==
2320 I1 = I + 1
2330 B(I) = B(I) - M1
2340 B(I1) = B(I1) - M2
2350   RETURN
2360   FOR I = 2 TO N
2370 I1 = I - 1
2380 B(I) = B(I) - A(I) * B(I1)
2390   NEXT I
2400 B(N) = B(N) / D(N)
2410   FOR I = N1 TO 1 STEP  - 1
2420 I1 = I + 1
2430 B(I) = (B(I) - C(I) * B(I1)) / D(I)
2440   NEXT I
2450   RETURN
2460   PRINT
```

```
2470    PRINT
2480    PRINT "SOLUTION FOR LOAD CASE ";K
2490    PRINT "****************************"
2500    PRINT
2505    REM ==OUTPUT JOINT ROTATIONS==
2510    GOSUB 2570
2520    PRINT "CONTINUE     (Y/N) ";
2530    INPUT C$
2540    IF C$ <  > "Y" THEN 2560
2550    GOSUB 2660
2560    RETURN
2570    PRINT
2580    PRINT "***JOINT ROTATIONS***"
2590    PRINT
2600    FOR I = 1 TO N
2610    PRINT "JOINT ";I;"  ROT = ";B(I)
2620    PRINT
2630    NEXT I
2640    PRINT
2650    RETURN
2660    PRINT
2670    PRINT
2680    PRINT "***MEMBER FORCES***"
2690 FOR I = 1 TO N1
2700 I1 = I + 1
2710 R1 = B(I)
2720 R2 = B(I1)
2730 S1 = K1(I)
2740 S2 = K2(I)
2750 S3 = K3(I)
2760 P1 = S1 * R1 + S3 * R2
2770 P2 = S3 * R1 + S2 * R2
2780 P1 = P1 + Q1(I)
2790 P2 = P2 + Q2(I)
2800    PRINT
2810    PRINT "> SPAN ";I
2820    PRINT
2830    PRINT "LEFT MOMENT = ";P1
2840    PRINT
2850    PRINT "RIGHT MOMENT = ";P2
2860    NEXT I
2870    RETURN
2880    END
```

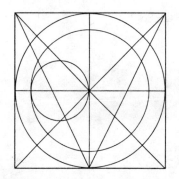

11
Thermodynamics and Heat Transfer

Thermodynamic Properties of Air

This program computes the thermodynamic properties of moist air for various conditions of barometric pressure, dry bulb temperature, and wet bulb temperature, based on the general format developed by the National Bureau of Standards.

Program Notes

The inputs required are:

- Barometric pressure in inches of mercury (29.921 standard for air)
- Dry bulb temperature (degrees Fahrenheit)
- Wet bulb temperature (degrees Fahrenheit)

Example

Compute the thermodynamic properties of air which has the following characteristics:

- Dry bulb temperature   80°F
- Wet bulb temperature   70°F
- Barometric pressure   29.921 in. of mercury

```
RUN
THERMODYNAMIC PROPERTIES OF AIR


BAR PRESSURE (INCHES HG)   = ?29.921
DRY BULB TEMP (DEG F)      = ?80
WET BULB TEMP (DEG F)      = ?70


DEW POINT TEMP (DEG F)     = 65.4128896
VAPOR PRESSURE (INCH HG)   = .631994357
HUMIDITY RATIO (#WAT/#AIR) = .0134214352
VOL MOIST AIR (FT^3/#)     = 13.8933352
ENTHALPY       (BTU/#)     = 33.9168721
REL HUMIDITY   (PERCENT)   = 61.2235887

RERUN PROGRAM WITH NEW DATA Y/N = ?N
```

Program Listing

```
90   HOME
100  PRINT "THERMODYNAMIC PROPERTIES OF AIR"
110  PRINT
120  PRINT
```

```
130   PRINT "BAR PRESSURE (INCHES HG)   = ";
140   INPUT P
150   PRINT "DRY BULB TEMP (DEG F)       = ";
160   INPUT T1
170   PRINT "WET BULB TEMP (DEG F)       = ";
180   INPUT T2
190   PRINT
200   PRINT
210   REM ==CHECK INPUT DATA==
220   IF T1 = > T2 THEN 270
240   PRINT "WET BULB MUST NOT EXCEED DRY BULB"
250   GOTO 150
260   REM ==CONVERT TO RANKIN==
270 T4 = (T2 + 459.688) / 1.8
280   REM ==CALCULATE SATURATION PRESSURE/WET BULB==
290   GOSUB 1020
300 P8 = P1
310   REM ==CONVERT TO RANKIN==
320 T4 = (T1 + 459.688) / 1.8
330   REM ==CALC SATURATION PRESSURE/DRY BULB==
340   GOSUB 1020
350 P9 = P1
370 W1 = 0.622 * P8 / (P - P8)
380   IF T2 < 32 THEN 450
390   IF T2 = 32 THEN 450
400 H1 = 1093.049 + 0.441 * T1 - T2
410 Z1 = 0.24 + 0.441 * W1
420 Z2 = W1 - Z1 * (T1 - T2) / H1
430 P2 = P * Z2 / (0.622 + Z2)
440   GOTO 470
450 P2 = P8 - 5.704E - 4 * P * (T1 - T2) / 1.8
460   REM ==CALC HUMIDITY RATIO W==
470 W = 0.622 * P2 / (P - P2)
480   REM ==CALC ENTHALPY H==
490 H = 0.24 * T1 + (1061 + 0.444 * T1) * W
500   REM  ==CALC DEW POINT T3==
510 T3 = T2
520   IF T1 = T2 THEN 590
530 L =  LOG (P2)
540   IF P2 < 0.18036 THEN 570
550 T3 = 79.047 + 30.579 * L + 1.8893 * L * L
560   GOTO 590
570 T3 = 71.98 + 24.873 * L + 0.8927 * L * L
580   REM ==CALC VOLUME MOIST AIR V==
590 V = 0.754 * (T1 + 459.7) * (1 + 7000 * W / 4360) / P
600   REM ==CALC RELATIVE HUMIDITY==
610 R = (P2 / P9) * 100
630   PRINT "DEW POINT TEMP (DEG F)      = ";T3
640   PRINT "VAPOR PRESSURE (INCH HG)    = ";P2
650   PRINT "HUMIDITY RATIO (#WAT/#AIR)= ";W
660   PRINT "VOL MOIST AIR   (FT^3/#)    = ";V
670   PRINT "ENTHALPY        (BTU/#)     = ";H
680   PRINT "REL HUMIDITY    (PERCENT)   = ";R
690   PRINT
700   PRINT "RERUN PROGRAM WITH NEW DATA Y/N = ";
```

```
710    INPUT Q$
720    IF Q$ = "Y" THEN 110
730    IF Q$ <  > "N" THEN 700
740    GOTO 1260
1000   REM   ==SUBROUTINE VAPOR PRESSURE OF SATURATED AIR==
1020 A1 =   - 7.90298
1030 A2 = 5.02808
1040 A3 =   - 1.3816E - 7
1050 A4 = 11.344
1060 A5 = 8.1328E - 3
1070 A6 =   - 3.49149
1080 B1 =   - 9.09718
1090 B2 =   - 3.56654
1100 B3 = 0.876793
1110 B4 = 0.0060273
1120   IF T4 < 273.16 THEN 1190
1130 Z = 373.16 / T4
1140 C1 = A1 * (Z - 1)
1150 C2 = A2 *  LOG (Z) /  LOG (10)
1160 C3 = A3 * (10 ^ (A4 * (1 - 1 / Z)) - 1)
1170 C4 = A5 * (10 ^ (A6 * (Z - 1)) - 1)
1180   GOTO 1240
1190 Z = 273.16 / T4
1200 C1 = B1 * (Z - 1)
1210 C2 = B2 *  LOG (Z) /  LOG (10)
1220 C3 = B3 * (1 - 1 / Z)
1230 C4 =  LOG (B4) /  LOG (10)
1240 P1 = 29.921 * (10 ^ (C1 + C2 + C3 + C4))
1250   RETURN
1260   END
```

Properties of Water

This program computes the density, specific gravity, and volumetric expansion of water at atmospheric pressure for various temperatures using a fifth order polynomial to calculate density and specific gravity.

Program Notes

The program may be run in either Fahrenheit or centigrade.

Example

Find the density and specific gravity of water at 60°F and at 20°C.

```
RUN
PROPERTIES OF WATER


FAHRENHEIT OR CENTIGRADE (F/C)
?F

INPUT INITIAL TEMP (32-212 F ) = ?60

DENSITY (KG/M^3) = 999.016447
DENSITY (#/FT^3) = 62.3665588
DENSITY (#/GAL)  = 8.33716158
SPECIFIC GRAVITY = .999041423

CALCULATE VOLUME EXPANSION ? Y/N = ?Y

INPUT FINAL TEMP (32-212 F ) = ?212

VOLUME OF WATER  V = ?12

DENSITY (KG/M^3) = 958.364856
DENSITY (#/FT^3) = 59.8287629
DENSITY (#/GAL)  = 7.99790902
SPECIFIC GRAVITY = .958388815

ORIGINAL VOLUME  = 12
FINAL VOLUME     = 12.5090119
DIFFERENCE VOL   = .509011887

RERUN PROGRAM WITH NEW DATA Y/N = ?N
```

Program Listing

```
90    HOME
100    PRINT "PROPERTIES OF WATER"
110    PRINT
120    PRINT
130    PRINT : PRINT "FAHRENHEIT OR CENTIGRADE (F/C)"
140    INPUT Q1$
150    PRINT
160    IF Q1$ = "F" THEN 240
170    IF Q1$ < > "C" THEN 130
180    PRINT "INPUT INITIAL TEMP (0-100 C ) = ";
190    INPUT C
200    IF C < 0 THEN 180
210    IF C > 100 THEN 180
220    PRINT
230    GOTO 310
240    PRINT "INPUT INITIAL TEMP (32-212 F ) = ";
250    INPUT F
260    IF F < 32 THEN 240
270    IF F > 212 THEN 240
280    PRINT
300 C = 5 * (F - 32) / 9
310    REM   CALC DENSITY
320    GOSUB 1000
330 D3 = D1
340    PRINT "CALCULATE VOLUME EXPANSION ? Y/N = ";
350    INPUT Q2$
360    PRINT
370    IF Q2$ = "N" THEN 660
380    IF Q2$ < > "Y" THEN 340
390    IF Q1$ = "F" THEN 460
400    PRINT "INPUT FINAL TEMP (0-100 C ) = ";
410    INPUT C
420    IF C < 0 THEN 400
430    IF C > 100 THEN 400
440    PRINT
450    GOTO 530
460    PRINT "INPUT FINAL TEMP (32-212 F ) = ";
470    INPUT F
480    IF F < 32 THEN 460
490    IF F > 212 THEN 460
500    PRINT
520 C = 5 * (F - 32) / 9
530    PRINT "VOLUME OF WATER  V = ";
540    INPUT V
550    PRINT
560    REM ==CALCULATE DENSITY==
570    GOSUB 1000
580 D4 = D1
590    REM ==CALCULATE VOLUME EXPANSION==
600 V1 = V * D3 / D4
610 V2 = V1 - V
620    PRINT "ORIGINAL VOLUME   = ";V
630    PRINT "FINAL VOLUME      = ";V1
```

```
640    PRINT "DIFFERENCE VOL    = ";V2
650    PRINT
660    PRINT "RERUN PROGRAM WITH NEW DATA Y/N = ";
670    INPUT Q3$
680    IF Q3$ = "Y" THEN  HOME : GOTO 120
690    IF Q3$ <  > "N" THEN 660
700    GOTO 1220
1010 A1 = 999.8425
1020 A2 = 16.9451784 * C
1030 A3 = 7.9870401E - 3 * C ^ 2
1040 A4 = 4.6170461E - 5 * C ^ 3
1050 A5 = 1.0556302E - 7 * C ^ 4
1060 A6 = 2.8054253E - 10 * C ^ 5
1070 A7 = .01687985 * C
1080 D1 = (A1 + A2 - A3 - A4 + A5 - A6) / (1 + A7)
1090   REM ==CONVERT TO #/FT^3==
1100 D2 = D1 * .06242796
1110   REM ==CALC #/GAL==
1120 G1 = D2 * .13368
1130   REM ==CALC SPECIFIC GRAVITY==
1140 S1 = D1 / 999.975
1160   PRINT "DENSITY (KG/M^3) = ";D1
1170   PRINT "DENSITY (#/FT^3) = ";D2
1180   PRINT "DENSITY (#/GAL)  = "G1
1190   PRINT "SPECIFIC GRAVITY = ";S1
1200   PRINT
1210   RETURN
1220   END
```

Saturation Steam Pressure

This program computes the saturation steam pressure as a function of temperature using the equation of Keenan, Keyes, Hill, and Moore. It will run using either Fahrenheit or centigrade and provides an output in saturation pressure measured in megapascals (MPa [10^7 dynes/cm^2]) and pounds per square inch absolute (PSIA).

Program Notes

The program is designed to run with the following parameters:

Maximum temperature
(°F) 705.44
(°C) 374.15

Minimum temperature
(°F) 32.00
(°C) 0.00

Example

Determine the operating pressure of a steam boiler at 220°F.

```
RUN
SATURATION STEAM PRESSURE


FAHRENHEIT OR CENTIGRADE F/C = ?F

INPUT TEMPERATURE (32-705 F) = ?220

PRESSURE - MPA    = .118506692
PRESSURE - PSIA   = 17.188085

RERUN WITH NEW DATA Y/N = ?N
```

Program Listing

```
90   HOME
100   PRINT "SATURATION STEAM PRESSURE"
120   DIM F(8)
130   PRINT
140   PRINT
150   PRINT "FAHRENHEIT OR CENTIGRADE F/C = ";
160   INPUT T$
```

```
170    PRINT
180    IF T$ = "F" THEN 230
190    IF T$ <  > "C" THEN 130
200    PRINT "INPUT TEMPERATURE (0-374 C) = ";
210    INPUT C
220    GOTO 270
230    PRINT "INPUT TEMPERATURE (32-705 F) = ";
240    INPUT F
250    REM ==CONVERT TO KELVIN==
260 C = 5 * (F - 32) / 9
270 K = C + 273.15
280 K1 = 1000 / K
290    REM ==CRITICAL TEMPERATURE/PRESSURE==
300 T1 = 374.136
310 P1 = 220.88
320    REM ==POLYNOMIAL COEFFICIENTS==
330 F(1) =  - 741.9242
340 F(2) =  - 29.72100
350 F(3) =  - 11.55286
360 F(4) =  - 0.8685635
370 F(5) = 0.1094098
380 F(6) = 0.439993
390 F(7) = 0.2520658
400 F(8) = 0.05218684
420 P = 0
430    FOR J = 1 TO 8
440 P = P + F(J) * (0.65 - .01 * C) ^ (J - 1)
450    NEXT J
460 P = P1 *  EXP (K1 * 1E - 5 * (T1 - C) * P) / 10
480 P2 = P * 145.03894
500    PRINT
510    PRINT "PRESSURE - MPA    = ";P
520    PRINT "PRESSURE - PSIA   = ";P2
530    PRINT
540    PRINT "RERUN WITH NEW DATA Y/N = ";
550    INPUT Q$
560    PRINT
570    IF Q$ = "Y" THEN 130
580    IF Q$ <  > "N" THEN 540
590    END
```

Barometric Pressure Correction

This program computes the barometric pressure (in inches of mercury) for various altitudes from 0 to 15,000 feet based on U.S. Standard Atmosphere. (The U.S. Standard Atmosphere is the barometric pressure which will support a column of mercury 29.921 inches high at sea level at a temperature of 59°F.)

Program Notes

While the barometric pressure of air varies considerably with altitude, temperature, and geographic location, this program assumes a linear decrease in temperature in the lower atmosphere. It also assumes that the lower atmosphere acts as a perfect gas, not taking into account the effects of geographic location.

Example

Compute the standard barometric pressure at sea level (0 feet altitude) and the corresponding standard barometric pressure for Denver, Colorado (5283 feet altitude).

```
RUN
BAROMETRIC PRESSURE CORRECTION

ALTITUDE = ?5283

BAROMETRIC PRESSURE

INCHES OF MERCURY  =24.6405843
MM OF MERCURY      =625.876508
POUNDS PER SQ INCH =12.1032111

TEMPERATURE FAHR.  =40.1638656

MORE DATA (Y/N)?N
```

Program Listing

```
90   HOME
100   PRINT "BAROMETRIC PRESSURE CORRECTION"
110   PRINT
120   PRINT "ALTITUDE = ";
130   INPUT A
140   PRINT
150   REM ==EQUATION COEFFICIENTS==
160 A1 = 29.921
170 A2 = 1.07097479E - 3
180 A3 = 1.35271553E - 8
190 B1 = 3.5654239E - 3
```

```
200  B2 = A1 - A2 * A + A3 * A * A
210  T = 59.0 - B1 * A
220  P = B2 * .4911901
230  M = B2 * 25.40023
240    PRINT "BAROMETRIC PRESSURE"
250    PRINT
260    PRINT "INCHES OF MERCURY   =";B2
270    PRINT "MM OF MERCURY       =";M
280    PRINT "POUNDS PER SQ INCH =";P
290    PRINT
300    PRINT "TEMPERATURE FAHR.   =";T
310    PRINT
320    PRINT "MORE DATA (Y/N)";
330    INPUT X$
340    IF X$ = "Y" THEN 110
350    IF X$ <  > "N" THEN 320
360    END
```

n5back Nav## 202ENT

Solar Intensity

This program computes the solar azimuth angle, solar altitude angle, and the intensity of the direct solar beam for any hour of a typical 365-day year. It also calculates the time of sunrise and sunset for that particular day.

The program is based on the format recommended by the American Society of Heating, Refrigeration and Air Conditioning Engineers.

Program Notes

This program does not take into account the one hour adjustment for daylight-saving time.

Example

Find the solar position and intensity for July 21 at 2:00 PM (14 hours past midnight) for an observer in time zone 8 at latitude 24° north and longitude 120° west.

```
RUN

SOLAR INTENSITY


INPUT MONTH (1-12)    = ?7
INPUT DAY   (1-31)    = ?21
INPUT HOUR  (0-23)    = ?14

EASTERN STANDARD   = 5
CENTRAL STANDARD   = 6
MOUNTAIN STANDARD  = 7
PACIFIC STANDARD   = 8

INPUT TIME ZONE NUMBER = ?8

INPUT LATITUDE  (DEGREES) = ?24
INPUT LONGITUDE (DEGREES) = ?120


DAY NUMBER       (1-365) = 202.583333
 TIME OF SUNRISE (HOURS) = 5.46942551
TIME OF SUNSET   (HOURS) = 18.5305745
DECLINATION ANGLE (DEG)  = 20.4260354
EQUATION OF TIME  (MIN)  = -6.34256236
SOLAR ALTITUDE (DEGREES) = 63.4981071
SOLAR AZIMUTH  (DEGREES) = 267.891428
SOLAR INTENSITY (BTU/FT^2)= 274.285491

RERUN PROGRAM WITH NEW DATA Y/N = ?N
```

Program Listing

```
100   HOME : PRINT
110   PRINT "SOLAR INTENSITY"
120   DIM M(12)
130   REM ==COMMON CONSTANTS==
140 P1 = 3.1415926
150 P2 = .0172021
160 P3 = 57.295778
170   REM ==LOAD CALENDAR DATA==
180   DATA  00,31,28,31,30,31
190   DATA  30,31,31,30,31,30
200   FOR J = 1 TO 12
210   READ M(J)
220   NEXT J
230   PRINT
240   PRINT
250   PRINT "INPUT MONTH  (1-12)     = ";
260   INPUT M1
270   PRINT "INPUT DAY    (1-31)     = ";
280   INPUT D
290   PRINT "INPUT HOUR   (0-23)     = ";
300   INPUT H
310   REM ==CONVERT TO DAY NUMBER==
320 N = 0
330   FOR J = 1 TO M1
340 N = N + M(J)
350   NEXT J
360 N = N + D + H / 24
370   PRINT
380   PRINT "EASTERN STANDARD   = 5"
390   PRINT "CENTRAL  STANDARD  = 6"
400   PRINT "MOUNTAIN STANDARD  = 7"
410   PRINT "PACIFIC STANDARD   = 8"
420   PRINT
430   PRINT "INPUT TIME ZONE NUMBER = ";
440   INPUT Z
450   PRINT
460   REM ==CALC DECLINATION==
470 D1 = .406079073
480 D2 = 1.40446929
490 D3 = .0132989724
500 D4 = 1.51396507
510 D5 = N * P2
520 R1 = D1 *  SIN (D5 - D2) + D3 * ( COS (D5 - D4) ^ 2)
530 A1 = R1 * P3
540   REM ==EQUATION OF TIME==
550 T1 = .16797268
560 T2 = .0520416347
570 T3 = 19.8500733
580 T4 = 7.40960196
590 T5 = N * P2 + T1
600 T6 = N * P2 - T2
610 T =  - T3 *  SIN (T5) *  COS (T5) - T4 *  SIN (T6)
620   PRINT "INPUT LATITUDE  (DEGREES) = ";
```

```
630    INPUT L1
640    PRINT "INPUT LONGITUDE (DEGREES) = ";
650    INPUT L2
660   REM ==CALC SUNRISE/SUNSET==
670 B1 =  -  TAN (L1 / P3) *  TAN (R1)
680 B2 =  -  ATN (B1 /  SQR ( - B1 * B1 + 1)) + P1 / 2
690 B3 = B2 * 12 / P1
700 B4 = 12 - B3 - (T / 60) - Z + L2 / 15
710 B5 = 24 - B4
720   REM  ==CALC HOUR ANGLE==
730 H1 = 15 * (H - 12 + Z + (T / 60)) - L2
740 H2 = H1 / P3
745   REM ==ANGLES OF SOLAR BEAM==
750 F4 = 0
760   IF  ABS (H2) >  ABS (B2) THEN 1030
780 C1 =  COS (L1 / P3)
790 C2 =  COS (R1)
800 C3 =  COS (H2)
810 S1 =  SIN (L1 / P3)
820 S2 =  SIN (R1)
830   REM  ==CALC SOLAR ALTITUDE==
840 S3 = C1 * C2 * C3 + S1 * S2
850 R2 =  ATN (S3 /  SQR ( - S3 * S3 + 1))
860 A2 = R2 * P3
880   REM ==CALC SOLAR AZIMUTH==
890 C4 = (S3 * S1 - S2) / ( COS (R2) * C1)
900 R3 =  -  ATN (C4 /  SQR ( - C4 * C4 + 1)) + P1 / 2
910 A3 =  ABS (R3 * P3)
920   IF H1 > 0 THEN 950
930 A3 = 180 - A3
940   GOTO 980
950 A3 = 180 + A3
970   REM ==CALC SOLAR INTENSITY==
980 F1 = N * .0172142
990 F2 = 368.4 + 24.538 *  COS (F1 - .02431)
1000 F3 = .17167 - .03475 *  COS
1010 F4 = F2 *  EXP ( - F3 / S3)
1020   REM ==PRINT RESULTS==
1030   PRINT
1040   PRINT
1050   PRINT "DAY NUMBER        (1-365) = ";N
1060   PRINT " TIME OF SUNRISE   (HOURS) = ";B4
1070   PRINT "TIME OF SUNSET     (HOURS) = ";B5
1080   PRINT "DECLINATION ANGLE (DEG)    = ";A1
1090   PRINT "EQUATION OF TIME   (MIN)   = ";T
1100   PRINT "SOLAR ALTITUDE (DEGREES)   = ";A2
1110   PRINT "SOLAR AZIMUTH  (DEGREES)   = ";A3
1120   PRINT "SOLAR  INTENSITY (BTU/FT^2)= ";F4
1130   PRINT
1140   PRINT "RERUN PROGRAM WITH NEW DATA Y/N = ";
1150   INPUT Q$
1160   IF Q$ = "Y" THEN 230
1170   IF Q$ <  > "N" THEN 1130
1180   END
```

12
Miscellaneous

Electric Motor Performance

Friction Loss in Pipe

Radioactive Decay

Acceleration Due to Gravity

Characteristic Equation of a Matrix

Electric Motor Performance

This program computes electric motor performance characteristics for standard open drip-proof 1-250 horsepower (1800 RPM, T-frame) three-phase induction motors.

Program Notes

The data tables at line 1300 contain specific values for nominal horsepower, full load amperage, and load power factor. This specific data is used to adjust the generalized performance curves.

The program may be tailored to a specific manufacturer simply by changing the data statements at the end of the program.

The program is designed to run with nominal voltage ranges as specified below:

Nominal Voltage	Minimum Voltage	Maximum Voltage
200/60/3	170	230
230/60/3	195.5	264.5
460/60/3	391	529

Example

Compute the operating characteristics of a nominal 10 hp motor wound for 460 volts when applied to a 7.5 hp load on a 460 volt line.

```
RUN
ELECTRIC MOTOR PERFORMANCE


INPUT NOMINAL HP (1-250) = ?10
NOM VOLTS (200,230,460)  = ?460
INPUT LOAD HORSEPOWER    = ?7.5
INPUT LINE VOLTAGE       = ?460


FULL LOAD HP      = 10
FL AMPERAGE       = 13.5
FL POWER FACTOR   = .805
FL KILOWATTS      = 8.65860859
FL EFFECIENCY     = .8615703

RATED LOAD HP     = 7.5
RL AMPERAGE       = 10.9353798
RL POWER FACTOR   = .742316918
RL KILOWATTS      = 6.4675782
RL EFFECIENCY     = .865084244

RERUN PROGRAM WITH NEW DATA Y/N = ?N
```

Program Listing

```
100    HOME
110    PRINT "ELECTRIC MOTOR PERFORMANCE"
120    DIM N(20),A(20),P(20)
140    REM ==READ MOTOR DATA==
150    FOR J = 1 TO 20
160    READ N(J),A(J),P(J)
170    NEXT J
180    PRINT
190    PRINT
200    PRINT "INPUT NOMINAL HP (1-250) = ";
210    INPUT H1
220    PRINT "NOM VOLTS (200,230,460)  = ";
230    INPUT V1
240    PRINT "INPUT LOAD HORSEPOWER    = ";
250    INPUT H2
260    PRINT "INPUT LINE VOLTAGE       = ";
270    INPUT V2
280    PRINT
290    PRINT
300    REM ==TEST INPUT DATA==
310    IF H1 < 1 THEN 440
320    IF H1 > 250 THEN 440
330    IF H2 > H1 THEN 440
340    IF V1 = 200 THEN 380
350    IF V1 = 230 THEN 380
360    IF V1 = 460 THEN 380
370    GOTO 480
380 Y = H2 / H1
390    IF Y < .5 THEN 440
400 Z = V2 / V1
410    IF Z < .85 THEN 460
420    IF Z > 1.15 THEN 460
430    GOTO 510
440    PRINT "INCORRECT NOMINAL HORSEPOWER"
450    GOTO 180
460    PRINT "OUTSIDE VOLTAGE UTILIZATION RANGE"
470    GOTO 180
480    PRINT "INCORRECT NOMINAL VOLTAGE"
490    GOTO 180
500    REM ==TEST FOR NOMINAL SIZE==
510 K = 1
520    IF N(K) >  = H1 THEN 560
530 K = K + 1
540    GOTO 520
550    REM ==CALCULATE FULL LOAD==
560 X = N(K)
570 Y = 1
580    GOSUB 1000
590 F1 = A(K) / B
600 F2 = P(K) / C
610 A1 = A(K) * 460 / V1
620    REM ==CALCULATE FULL LOAD==
630 Y = H2 / H1
```

```
640    GOSUB 1000
650 A2 = B * 460 / V1
660 P1 = C
670   REM ==CALCULATE VOLTAGE ADJUSTMENT==
680 F3 = 1
690 F4 = 1
700   IF Z = 1 THEN 740
710 F3 = 7.17311269 - 11.9961805 * Z + 5.82460674 * Z * Z
720 F4 =  - .765600727 + 4.28767874 * Z - 2.52543 * Z * Z
730   REM ==CALCULATE MOTOR VALUES==
740 A3 = A2 * F1 * F3
750 P2 = P1 * F2 * F4
760 K1 = (V2 * A1 * F3 * P(K) *  SQR (3)) / 1000
770 E1 = (.746 * H1) / K1
780 K2 = (V2 * A3 * P2 *  SQR (3)) / 1000
790 E2 = (.746 * H2) / K2
810   PRINT "FULL LOAD HP     = ";H1
820   PRINT "FL AMPERAGE      = ";A1
830   PRINT "FL POWER FACTOR = ";P(K)
840   PRINT "FL KILOWATTS     = ";K1
850   PRINT "FL EFFICIENCY    = ";E1
860   PRINT
870   PRINT "RATED LOAD HP    = ";H2
880   PRINT "RL AMPERAGE      = ";A3
890   PRINT "RL POWER FACTOR = ";P2
900   PRINT "RL KILOWATTS     = ";K2
910   PRINT "RL EFFICIENCY    = ";E2
920   PRINT
930   PRINT "RERUN PROGRAM WITH NEW DATA Y/N = ";
940   INPUT Q$
950   IF Q$ = "Y" THEN 180
960   IF Q$ <  > "N" THEN 920
970   GOTO 1520
1000   REM ==POLYNOMIAL-AMPS==
1010 B1 = 3.20607431
1020 B2 = .310338345 * X
1030 B3 =  - 3.35324861E - 4 * X ^ 2
1040 B4 =  - .506590351 * Y
1050 B5 = .155589858 * Y ^ 2
1060 B6 = .4725898 * X * Y
1070 B7 = 6.77369891E - 4 * X ^ 2 * Y
1080 B8 = .366744037 * X * Y ^ 2
1090 B9 =  - 4.6061703E - 4 * X ^ 2 * Y ^ 2
1100 B = B1 + B2 + B3 + B4 + B5 + B6 + B7 + B8 + B9
1110   REM ==POLYNOMIAL EFFICIENCY==
1120 C1 = .309145751
1130 C2 = 1.23049012E - 3 * X
1140 C3 =  - 3.76440456E - 7 * X ^ 2
1150 C4 = .725339116 * Y
1160 C5 =  - .25574737 * Y ^ 2
1170 C6 = 4.22942991E - 3 * X * Y
1180 C7 =  - 2.04667461E - 5 * X ^ 2 * Y
1190 C8 =  - 4.28755616E - 3 * X * Y ^ 2
1200 C9 = 1.77828477E - 5 * X ^ 2 * Y ^ 2
1210 C = C1 + C2 + C3 + C4 + C5 + C6 + C7 + C8 + C9
```

```
1220    RETURN
1300    REM  ==FULL LOAD MOTOR DATA==
1310    REM     HP,AMPS,PF
1320    DATA    1,1.7,.710
1330    DATA    1.5,2.82,.673
1340    DATA    2,3.15,.770
1350    DATA    3,4.6,.777
1360    DATA    5,7.0,.799
1370    DATA    7.5,10.5,.780
1380    DATA    10,13.5,.805
1390    DATA    15,20,.809
1400    DATA    20,26,.822
1410    DATA    25,32,.818
1420    DATA    30,37,.841
1430    DATA    40,50,.831
1440    DATA    50,62,.829
1450    DATA    60,74,.840
1460    DATA    75,90,.857
1470    DATA    100,120,.856
1480    DATA    125,148,.862
1490    DATA    150,174,.875
1500    DATA    200,228,.883
1510    DATA    250,281,.893
1520    END
```

Friction Loss in Pipe

This program computes the friction loss of water flowing in various pipes using the empirical formula developed by Williams and Hazen. The program also computes water velocity, flow in gallons, and weight of the water and pipe.

The empirical formula used in this program is:

$$F = 0.2083 \times \frac{100}{C} \times G^{1.852} \times L$$

$$100 \times D^{4.0655}$$

where:

F = friction loss in feet of water
C = water flow in gallons
G = water flow in gallons per minute
D = inside diameter of pipe in inches
L = length of pipe in feet

Example

Compute the friction loss for 20 gallons per minute of water flowing in a 1 inch, type L copper pipe that is 100 feet long.

```
RUN
FRICTION LOSS IN PIPE

WHAT TYPE OF PIPE -

1 - COPPER TYPE L
2 - STEEL SCHEDULE 40
3 - PVC PLASTIC SCHEDULE-40
4 - PVC PLASTIC SCHEDULE-80
5 - COPPER TYPE K

INPUT 1-4 = ?1

WATER FLOW (GALLONS/MINUTE)   = ?20
TOTAL LENGTH OF PIPE (FEET)   = ?100
NOMINAL PIPE SIZE (.5-8 INCH) = ?1

NOMINAL PIPE SIZE - INCHES = 1
OUTSIDE DIAMETER  - INCHES = 1.125
INSIDE   DIAMETER  - INCHES = 1.025

FRICTION LOSS- FT OF WATER = 23.2352928
FRICTION LOSS- #/SQ INCH   = 10.0585683

VELOCITY OF WATER (FT/SEC) = 7.77682621
GALLONS OF WATER IN PIPE   = 4.28617539
```

```
WEIGHT OF WATER    (POUNDS) = 35.6995548
WEIGHT OF PIPE     (POUNDS) = 65.5155138
WEIGHT OF WATER AND PIPE    = 101.215069

RE-RUN PROGRAM WITH NEW DATA Y/N = ?N
```

Program Listing

```
10    HOME
100    PRINT "FRICTION LOSS IN PIPE"
110    PRINT
120    DIM C(5),W(5),N1(12)
130    DIM D2(5,12),D1(5,12)
140    REM ==READ PIPE DATA==
150    FOR I = 1 TO 12
160    READ N1(I)
170    NEXT I
180    FOR K = 1 TO 5
190    READ C(K)
200    READ W(K)
210    FOR J = 1 TO 12
220    READ D2(K,J)
230    READ D1(K,J)
240    NEXT J
250    NEXT K
260    PRINT "WHAT TYPE OF PIPE - "
270    PRINT
280    PRINT "1 - COPPER TYPE L"
290    PRINT "2 - STEEL SCHEDULE 40"
300    PRINT "3 - PVC PLASTIC SCHEDULE-40 "
310    PRINT "4 - PVC PLASTIC SCHEDULE-80 "
315    PRINT "5 - COPPER TYPE K"
320    PRINT
330    PRINT "INPUT 1-4 = ";
340    INPUT K
350    PRINT
360    PRINT "WATER FLOW (GALLONS/MINUTE)   = ";
370    INPUT G
380    PRINT "TOTAL LENGTH OF PIPE (FEET)   = ";
390    INPUT L
400    PRINT "NOMINAL PIPE SIZE (.5-8 INCH) = ";
410    INPUT N
420    PRINT
430    REM ==CHECK PIPE SIZE==
440    IF N > 8 THEN 400
450 J = 1
460    IF N1(J) = N THEN 500
470    IF N1(J) > N THEN 500
480 J = J + 1
490    GOTO 460
500    REM ==CALC PIPE CROSS SECTIONS==
510 A2 = (D2(K,J) / 24) ^ 2 * 3.14159
520 A1 = (D1(K,J) / 24) ^ 2 * 3.14159
530    REM ==CALC FRICTION LOSS==
```

```
540 F = .2083 * (100 * G / C(K)) ^ 1.852 * (L / 100) / D1(K,J) ^ 4.8655
550 P = F / 2.31
560   REM ==CALC WATER VELOCITY==
570 V = G / (60 * 7.48 * A1)
580   REM ==CALC WEIGHT OF WATER==
590 W1 = A1 * L * 62.3
600   REM ==CALC WEIGHT OF PIPE==
610 W2 = (A2 - A1) * L * W(K)
620   REM ==CALC GAL WATER IN PIPE==
630 G2 = W1 / 8.329
640   REM ==CALC WEIGHT WATER & PIPE==
650 W3 = W2 + W1
660   REM ===PRINT RESULTS==
670   PRINT "NOMINAL PIPE SIZE - INCHES = ";N1(J)
680   PRINT "OUTSIDE DIAMETER   - INCHES = ";D2(K,J)
690   PRINT "INSIDE  DIAMETER  - INCHES = ";D1(K,J)
700   PRINT
710   PRINT "FRICTION LOSS- FT OF WATER = ";F
720   PRINT "FRICTION LOSS- #/SQ INCH   = ";P
730   PRINT
740   PRINT "VELOCITY OF WATER (FT/SEC) = ";V
750   PRINT "GALLONS OF WATER IN PIPE   = ";G2
760   PRINT "WEIGHT OF WATER   (POUNDS) = ";W1
770   PRINT "WEIGHT OF PIPE    (POUNDS) = ";W2
780   PRINT "WEIGHT OF WATER AND PIPE   = ";W3
790   PRINT
800   PRINT "RE-RUN PROGRAM WITH NEW DATA Y/N = ";
810   INPUT Z$
820   IF Z$ = "Y" THEN 260
830   IF Z$ < > "N" THEN 790
840   GOTO 1430
1000   REM ==DATA NOMINAL PIPE==
1010   DATA  .5,.75,1,1.25,1.5,2
1020   DATA  2.5,3,4,5,6,8
1030   REM ==DATA COPPER PIPE TYPE L==
1040   DATA  147,558.7
1050   DATA  0.625,0.545,0.875,0.785
1060   DATA  1.125,1.025,1.375,1.265
1070   DATA  1.625,1.505,2.125,1.985
1080   DATA  2.625,2.465,3.125,2.945
1090   DATA  4.125,3.905,5.125,4.875
1100   DATA  6.125,5.845,8.125,7.725
1110   REM ==DATA STEEL PIPE SCH 40==
1120   DATA  135,488.8
1130   DATA  0.840,0.622,1.050,0.824
1140   DATA  1.315,1.049,1.660,1.380
1150   DATA  1.900,1.610,2.375,2.067
1160   DATA  2.875,2.469,3.500,3.068
1170   DATA  4.500,4.026,5.563,5.047
1180   DATA  6.625,6.065,8.625,7.981
1190   REM ==DATA PVC SCHED-40==
1200   DATA  150,86.6
1210   DATA  0.840,0.622,1.050,0.824
1220   DATA  1.315,1.049,1.660,1.380
1230   DATA  1.900,1.610,2.375,2.067
```

```
1240    DATA    2.875,2.469,3.500,3.068
1250    DATA    4.500,4.026,5.563,5.047
1260    DATA    6.625,6.065,8.625,7.981
1270    REM    DATA PVC SCHED-80
1279    REM  ==DATA PVC SCHED-80==
1280    DATA    150,86.6
1290    DATA    0.840,0.546,1.050,0.742
1300    DATA    1.315,0.957,1.660,1.278
1310    DATA    1.900,1.500,2.375,1.939
1320    DATA    2.875,2.323,3.500,2.900
1330    DATA    4.500,3.826,5.563,5.312
1340    DATA    6.625,5.761,8.625,7.625
1350    REM  ==COPPER TYPE K==
1360    DATA    147,558.7
1370    DATA    0.625,0.527,0.875,0.745
1380    DATA    1.125,0.995,1.375,1.245
1390    DATA    1.625,1.481,2.125,1.959
1400    DATA    2.625,2.435,3.125,2.907
1410    DATA    4.125,3.875,5.125,4.805
1420    DATA    6.225,5.741,8.125,7.583
1430    END
```

Radioactive Decay

This program computes the rate of decay of a radioactive isotope over a specified period of time. The rate of decay proceeds according to an exponential curve given by the formula:

$$N = N_1 \times E^{-KT}$$

where:

N = the amount of isotope remaining after time (T)

N_1 = the amount of isotope originally present

K = exponential constant

T = elapsed time

Program Notes

This program will solve for the following:

- Amount of radioactive decay in a given time
- Time to decay a given amount
- Carbon-14 dating problem

Example

A mummy has been unearthed in which the ratio of carbon-14 is 45% of that found in the atmosphere. What is the approximate age of the mummy?

```
RUN
RADIOACTIVE DECAY


DO YOU WISH TO SOLVE FOR

1 - AMOUNT OF DECAY IN A GIVEN TIME
2 - TIME TO DECAY A GIVEN AMOUNT
3 - CARBON-14 DATING PROBLEM

INPUT 1-2-3 ?3

PERCENT C-14 IN SAMPLE (1-100) = ?45

APPROXIMATE AGE IN YEARS = 6600


RE-RUN PROGRAM WITH NEW DATA Y/N = ?Y


DO YOU WISH TO SOLVE FOR
```

```
1 - AMOUNT OF DECAY IN A GIVEN TIME
2 - TIME TO DECAY A GIVEN AMOUNT
3 - CARBON-14 DATING PROBLEM

INPUT 1-2-3 ?2

INPUT RADIOACTIVE HALF-LIFE = ?2
PERCENT TO REMAIN (1-100) = ?35.35

TIME PERIOD OF DECAY = 3.00043576

RE-RUN PROGRAM WITH NEW DATA Y/N = ?Y


DO YOU WISH TO SOLVE FOR

1 - AMOUNT OF DECAY IN A GIVEN TIME
2 - TIME TO DECAY A GIVEN AMOUNT
3 - CARBON-14 DATING PROBLEM

INPUT 1-2-3 ?1

INPUT AMOUNT OF ORIGINAL SAMPLE = ?1
INPUT HALF-LIFE FROM TABLES = ?2
TIME PERIOD OF DECAY = ?3


AMOUNT OF REMAINING SAMPLE = .353553391
PERCENT OF REMAINING SAMPLE = 35.3553391

RE-RUN PROGRAM WITH NEW DATA Y/N = ?N
```

Program Listing

```
100   HOME
110   PRINT "RADIOACTIVE DECAY"
120   PRINT : PRINT
130   PRINT "DO YOU WISH TO SOLVE FOR"
140   PRINT
150   PRINT "1 - AMOUNT OF DECAY IN A GIVEN TIME"
160   PRINT "2 - TIME TO DECAY A GIVEN AMOUNT"
170   PRINT "3 - CARBON-14 DATING PROBLEM"
180   PRINT
190   PRINT "INPUT 1-2-3 ";
200   INPUT Z
210   PRINT
220   IF Z = 2 THEN 420
230   IF Z = 3 THEN 530
240   IF Z <  > 1 THEN 130
250   REM ==SOLVE FOR CASE 1==
260   PRINT "INPUT AMOUNT OF ORIGINAL SAMPLE = ";
270   INPUT N1
280   PRINT "INPUT HALF-LIFE FROM TABLES = ";
290   INPUT H
```

```
300   PRINT "TIME PERIOD OF DECAY = ";
310   INPUT T
320   PRINT
330   REM ==SOLVE EQUATION==
340 N = N1 * (.5 ^ (T / H))
350   PRINT
360   PRINT "AMOUNT OF REMAINING SAMPLE = "N
370 N2 = (N / N1) * 100
380   PRINT "PERCENT OF REMAINING SAMPLE = ";N2
390   PRINT
400   GOTO 610
410   REM ==SOLVE FOR CASE 2==
420   PRINT "INPUT RADIOACTIVE HALF-LIFE = ";
430   INPUT H
440   PRINT "PERCENT TO REMAIN (1-100) = ";
450   INPUT P
460   PRINT
470 K =  LOG (2) / H
480 T =  -  LOG (P / 100) / K
490   PRINT "TIME PERIOD OF DECAY = "T
500   PRINT
510   GOTO 610
520   REM ==SOLVE FOR CASE 3==
530   PRINT "PERCENT C-14 IN SAMPLE (1-100) = ";
540   INPUT R
550   PRINT
560 T = 5730 *  ABS ( LOG (R / 100)) /  LOG (2)
570 T =   INT (T)
580   PRINT "APPROXIMATE AGE IN YEARS = ";T
590   PRINT
600   PRINT
610   PRINT "RE-RUN PROGRAM WITH NEW DATA Y/N = ";
620   INPUT Q$
630   IF Q$ = "Y" THEN 120
640   IF Q$ <  > "N" THEN 600
650   END
```

Acceleration Due to Gravity

This program computes the acceleration due to gravity for any latitude and elevation using Helmert's approximation:

$$G = 980.616 - 2.5928 \cos(2L) + 0.0068 \cos^2(2L) - 1.0125 \times 10 - 5H$$

The International Committee on Weights and Measures has adopted a standard accepted value for the acceleration due to gravity as:

$$32.174 \text{ Ft/Sec} \times \text{Sec} \ (980.665 \text{ Cm/Sec} \times \text{Sec})$$

Example

Denver, Colorado is at 39°N latitude and 5280 feet elevation. Key West, Florida is a 24°N latitude and at sea level. Compare the acceleration due to gravity for the two locations.

```
RUN
ACCELERATION DUE TO GRAVITY


LATITUDE (0-90 DEG) = ?39
ALTITUDE (FEET)     = ?5280

G (FT/SEC*SEC) = 32.1530107
G (CM/SEC*SEC) = 980.023766

CONTINUE Y/N = ?Y



LATITUDE (0-90 DEG) = ?24
ALTITUDE (FEET)     = ?0

G (FT/SEC*SEC) = 32.1156223
G (CM/SEC*SEC) = 978.884168

CONTINUE Y/N = ?N
```

Program Listing

```
100   HOME
110   PRINT "ACCELERATION DUE TO GRAVITY"
120   PRINT : PRINT
130   PRINT "LATITUDE (0-90 DEG) = ";
140   INPUT L
150   PRINT "ALTITUDE (FEET)     = ";
160   INPUT H
170   PRINT
```

```
180   REM ===HELMERTS EQUATION===
190 R = 57.29577
200 A =  COS (2 * L / R)
210 A1 = 980.616
220 A2 = 2.5928 * A
230 A3 = .0069 * A * A
240 A4 = 1.0125E - 5 * H
250 G = A1 - A2 + A3 - A4
260   PRINT "G (FT/SEC*SEC) = "G / 30.48
270   PRINT "G (CM/SEC*SEC) = "G
280   PRINT
290   PRINT "CONTINUE Y/N = ";
300   INPUT Q$
310   PRINT
320   IF Q$ = "Y" THEN 120
330   IF Q$ <  > "N" THEN 290
340   END
```

Characteristic Equation of a Matrix

This program takes as its data the dimensions and elements of a matrix and generates the p's and performs the matrix multiplications. (Note from the function in line 60 that the p's are preceded by a minus sign.) The method provides a stable output for a stable input. For example, given a matrix A, whose elements are all integers, the p's which are generated should all be integers as well.

Example

$$\text{Given matrix } A = \begin{bmatrix} 11 & 6 & -2 \\ -2 & 18 & 1 \\ -12 & 24 & 13 \end{bmatrix}$$

```
CHARACTERISTIC EQUATION OF A MATRIX


KEY IN ORDER  3
          KEY IN ELEMENTS AS INDICATED

    A(1,1) =?11
    A(1,2) =?6
    A(1,3) =?-2

    A(2,1) =?-2
    A(2,2) =?18
    A(2,3) =?1

    A(3,1) =?-12
    A(3,2) =?24
    A(3,3) =?13

      <<<<< COEFFICIENTS ARE >>>>>

              P(1) = 42
              P(2) = -539
              P(3) = 2058
```

$$\text{thus } f(\lambda) = (-1)^3(L^3 - 42L^2 + 539L - 2058)$$

Program Listing

```
10   HOME
20   DIM A(10,10),B(10,10),P(10),C1(10)
30   PRINT "CHARACTERISTIC EQUATION OF A MATRIX"
40   PRINT
60   PRINT " F(LAMBDA) = (-1) * (L - P(1) L - P(2) L - ... - P(N))"
70   PRINT
```

```
80    PRINT : INPUT "KEY IN ORDER   ";N
90    PRINT "              KEY IN ELEMENTS AS INDICATED": PRINT
100   FOR I = 1 TO N
110   FOR J = 1 TO N
120   PRINT "      A(";I;",";J;") =";
130   INPUT A(I,J)
140  B(I,J) = A(I,J)
150   NEXT J
160   PRINT
170   NEXT I
180  M = N - 1
190   FOR K = 1 TO M
200  T1 = 0
210   FOR I = 1 TO N
220  T1 = T1 + B(I,I)
230   NEXT I
240  A1 = K
250  P(K) = T1 / A1
260   FOR I = 1 TO N
270  B(I,I) = B(I,I) - P(K)
280   NEXT I
290   FOR J = 1 TO N
300   FOR I = 1 TO N
310  C1(I) = B(I,J)
320   NEXT I
330   FOR I = 1 TO N
340  B(I,J) = 0
350   FOR L = 1 TO N
360  B(I,J) = B(I,J) + A(I,L) * C1(L)
370   NEXT L
380   NEXT I
390   NEXT J
400   NEXT K
410  P(N) = B(N,N)
420   HOME
430   PRINT "      <<<<< COEFFICIENTS ARE >>>>>"
440   PRINT
450   FOR K = 1 TO N
460   PRINT  TAB( 25)"P(";K;") = ";P(K)
470   NEXT K
480   END
```

Appendix

Appendix

There have been attempts in recent years to establish a standard for BASIC. However, no such standard is in widespread use, and this state of affairs is likely to continue indefinitely. In this book we have tried to use a generalized dialect of BASIC that is compatible with many different computers, but since they are written to run on the Apple II[1] you may find that some of the conventions we have adopted will not work on other computers. This appendix describes some of the changes you can make to the programs in this book to resolve BASIC compatibility problems. If you encounter problems that are not described here, you will have to use your own ingenuity to resolve them.

In this appendix you will also find suggestions for changing the programs to accommodate different output devices.

We describe each of the changes listed below in a general way. You must decide how a suggested change would apply to a particular program, if at all. You will therefore need some understanding of BASIC programming in order to implement these changes.

Pausing with Full Display Screen

Many programs have more lines of output than will fit on a typical display screen. This means the first lines of output flash by quickly and scroll off the top of the screen, leaving you with no idea of what they contained. On many computers, you can press a key or a combination of keys to temporarily freeze the display. You can then review and record anything on the display. Subsequently pressing the special key or combination of keys again sets the computer in motion. More program output appears. You may have to freeze the display several times in order to see all the output. The number of times you must freeze the display depends not only on which program you are running, but also on the nature of the problem you present it with. For example, in CBASIC™ you can simultaneously press the CONTROL and S keys to induce this state of suspended animation. On the Radio Shack TRS-80, you do the same thing by simultaneously pressing the SHIFT and @ keys.

Alternatively, you can modify a program so it pauses at one or more points during its output, waiting for the user to cue it to continue. To do this, add the following subroutine to the program, and call the subroutine at suitable intervals during the output phase of the program.

```
5799 REM WAIT FOR OPERATOR CUE
5800 PRINT "ENTER 'C' TO CONTINUE"
5810 INPUT W$
5820 RETURN
```

String Variables, Functions, and Concatenation

Programs which use simple (non-array) string variables do not dimension them in a DIM statement, in keeping with common practice. If your computer requires it, add a DIM statement to dimension such string variables at the beginning of each program that uses them. For example:

```
20 DIM D(12), 10(3,2), C0(4,3), C1(5,5), C1$(5) E0(25,2)
```

If your BASIC has different syntax rules for dimensioning string arrays, you may change the line accordingly.

---

1. Apple II is a trademark of Apple Computer, Inc.

Some programs use the string function MID$ to specify part of a string. Your BASIC may have a different name for this function, like SEG$. If so, replace each occurrence of MID$ with the name of the corresponding function in your BASIC.

A plus sign ($+$) is the most common BASIC concatenation operator. If your BASIC uses a different one, like an ampersand ($\&$), replace each occurrence of the plus sign (where it is used as a concatenation operator) with the correct concatenation operator. Be sure you do not replace any plus signs used for arithmetic addition with the concatenation operator.

Printer Output

Viewing program output on the display screen is perfectly acceptable when you are using a program as an experimental or investigative tool. But sooner or later you will probably tire of continually copying program output from the display by hand. The solution, of course, is to direct program output to a printer. The procedure for doing this varies widely from one computer to the next. In some cases you need only press a key or a combination of keys and all program output appears on both the display and the printer simultaneously. In other cases you can achieve this same effect (or cause output to appear only on the printer) by entering a sequence of commands just before you run a program.

In many cases, though, you must actually modify each PRINT statement in a program in order to reroute output to the printer. Here again, the procedure for doing this varies widely. Consult your BASIC reference manual for specific instructions.

Replacing DEF FN Statements

Most BASICs allow you to define your own single-line function by using a DEF FN statement. You invoke such functions with a FN statement. For example:

$$20 \text{ DEF FNA}(X) = \text{INT}(X \times 100 + 0.5)/100$$
$$240 \text{ T} = \text{T} + \text{FNA}(C(J) \times (1 + R)!(N-J))$$

You can replace such user-defined functions with regular subroutines. The DEF FN statement becomes the subroutine and must be placed outside the normal mainline execution path of the program. In each place the function is invoked with an FN statement, you must insert a call to the new subroutine:

$$1000 \text{ X} = \text{INT}(X \times 3100 + 0.5)/100$$
$$1010 \text{ RETURN}$$

You would change line 240 as shown below, and add new lines to call the new subroutine.

$$240 \text{ X}=C(J) \times (1+R)!(N-J)$$
$$243 \text{ GOSUB } 1000$$
$$245 \text{ T} = \text{T} + \text{X}$$

The Authors

Richard Bennington is a graduate of Miami University. He is currently an engineer at the Borg-Warner Corporation. Contribution:

- Regression of an Arbitrary Function in One Variable
- Thermodynamic Properties of Air
- Properties of Water
- Saturation Steam Pressure
- Barometric Pressure Correction
- Solar Intensity
- Electric Motor Performance
- Friction Loss in Pipe
- Radioactive Decay
- Acceleration Due to Gravity

Marcial Blondet completed a master's degree in Civil Engineering at the University of California in Berkeley, where he is currently at work on his doctorate.

Juan Carlos Simo completed his undergraduate work at the Universidad Politecnica in Madrid, earning the degree Ingeniero Superior. As a Fulbright fellow, he completed a master's degree in Civil Engineering at the University of California in Berkeley, where he is currently at work on his dissertation.

Together they contributed the following:

- Fourier Series Analysis of Periodic Functions
- Fourier Series Expansion of Piece-Wise Linear Periodic Functions
- State of Stresses at a Point
- Geometric Properties of an Arbitrary Plane Domain
- Bending Moments and Shear Force Envelopes
- Fixed End Moments and Stiffness Coefficients for Beams
- Joint Rotations and End Moments for a Continuous Beam
- Compact Crout Method
- General Linear Equation Solver
- Inverse Iteration with Shifts
- Cyclic Jacobi Method
- Runge-Kutta Method for a First Order Equation
- Predictor-Corrector Method for a System of Equations
- Runge-Kutta Method for a System of Equations

Kenneth Douglass is an associate professor of Nuclear Medicine at Johns Hopkins where he is at work on microcomputer image processing systems for nuclear medicine. He received his undergraduate degree from the University of Rochester, and the master's and Ph.D. in Physics from Carnegie-Mellon University. Contribution:

- Peak Finder
- Data Bounding, Smoothing and Differentiation
- Convolution/Deconvolution
- Polynomial Regression in Two Variables
- Cubic Spline Interpolation

William Harlow is an associate professor of Mechanical and Industrial Engineering at the University of Cincinnati. He completed his undergraduate work at Miami University in Ohio, and earned a master's degree in Mechanical Engineering at the University of Cincinnati. Contribution:

- Integral Evaluation by the Modified Simpson's Method
- Gaussian Evaluation of a Definite Integral
- Fredholm Integral Equation
- Characteristic Equation of a Matrix
- Roots of the General Polynomial
- Eigenvalues of the General Matrix

Members of the Osborne/McGraw-Hill staff, Mary Borchers, Steven Cook, Lon Poole and Martin McNiff, contributed:

- Gaussian Elimination Method
- Multiple Linear Regression
- Alphabetize
- Roots of Polynomials: Half-Interval Search
- Real Roots of Polynomials: Newton
- Roots of Polynomials: Bairstow's Method
- Lagrangian Interpolation
- Linear Regression
- Geometric Regression
- Exponential Regression
- N th-Order Regression

Shafi Motiwalla is a research and development engineer who is currently involved in the analysis and design of microprocessor-controlled mechanical systems. He received a bachelor's degree in Mechanical Engineering from the Massachusetts Institute of Technology, and a master's degree in Mechanical Engineering and Computer Science from the University of California in Berkeley. Mr. Mortiwalla provided a technical review of the programs.

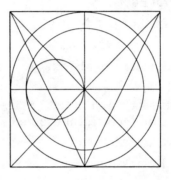

Other OSBORNE/McGraw-Hill Publications

An Introduction to Microcomputers: Volume 0 — The Beginner's Book
An Introduction to Microcomputers: Volume 1 — Basic Concepts, second edition
An Introduction to Microcomputers: Volume 2 — Some Real Microprocessors
An Introduction to Microcomputers: Volume 3 — Some Real Support Devices
Osborne 4 & 8-Bit Microprocessor Handbook
Osborne 16-Bit Microprocessor Handbook
8089 I/O Processor Handbook
CRT Controller Handbook
68000 Microprocessor Handbook
8080A/8085 Assembly Language Programming
6800 Assembly Language Programming
Z80 Asembly Language Programming
6502 Assembly Language Programming
Z8000 Assembly Language Programming
6809 Assembly Language Programming
Running Wild — The Next Industrial Revolution
The 8086 Book
PET and the IEEE 488 Bus (GPIB)
PET/CBM Personal Computer Guide, 2nd Edition
Business System Buyer's Guide
OSBORNE CP/M® User Guide
Apple II® User's Guide
Microprocessors for Measurement & Control
Some Common BASIC Programs
Some Common BASIC Programs — PET/CBM Edition
some Common BASIC Programs — TRS-80™ Level II Edition
Practical BASIC Programs
Payroll with Cost Accounting
Accounts Payable and Accounts Receivable
General Ledger
8080 Programming for Logic Design
6800 Programming for Logic Design
Z80 Programming for Logic Design